Herbs and Spices - Innovations in Extraction, Chemical Composition, Pharmacology, and Biotechnological Applications

Edited by Mozaniel Santana de Oliveira, Eloisa Helena De Aguiar Andrade, Hayet Edziri, Ravendra Kumar and Enéas Andrade Fontes Júnior

Published in London, United Kingdom

Herbs and Spices - Innovations in Extraction, Chemical Composition, Pharmacology, and Biotechnological Applications
http://dx.doi.org/10.5772/intechopen.1008098
Edited by Mozaniel Santana de Oliveira, Eloisa Helena De Aguiar Andrade, Hayet Edziri, Ravendra Kumar and Enéas Andrade Fontes Júnior

Contributors
Abdullah R. Alanzi, Ali Irfan, Bakar Bin Khatab Abbasi, Caroline da Costa Callandrine, Eloisa Helena de Aguiar Andrade, Fábio José Bonfim Cardoso, Hulme Ríos-Guerra, Huma Hameed, Jair Alejandro Temis-Cortina, Judith Espinosa-Raya, Karyme do Socorro de Souza Vilhena, Mahwish Arshad, Marcelo Alves Farias, Mozaniel Santana de Oliveira, Oberdan Oliveira Ferreira, Percin Karakol, Raquel Gómez-Pliego, Riaz Ullah, Sadaf Ahmad, Shazia Perveen, Sucheewin Krobthong, Tehreem Mushtaq, Yodying Yingchutrakul, Yousef A. Bin Jardan

First published in London, United Kingdom, 2026 by IntechOpen
IntechOpen is the global imprint of INTECHOPEN LIMITED, registered in England and Wales, registration number: 11086078, 167-169 Great Portland Street, London, W1W 5PF, United Kingdom

For EU product safety concerns: IN TECH d.o.o., Prolaz Marije Krucifikse Kozulić 3, 51000 Rijeka, Croatia, info@intechopen.com or visit our website at intechopen.com.

British Library Cataloguing-in-Publication Data
A catalogue record for this book is available from the British Library

Herbs and Spices - Innovations in Extraction, Chemical Composition, Pharmacology, and Biotechnological Applications
Edited by Mozaniel Santana de Oliveira, Eloisa Helena De Aguiar Andrade, Hayet Edziri, Ravendra Kumar and Enéas Andrade Fontes Júnior
p. cm.
Print ISBN 978-1-83634-959-4
Online ISBN 978-1-83634-958-7
eBook (PDF) ISBN 978-1-83634-960-0

Meet the editors

Dr. Mozaniel Santana de Oliveira holds a Bachelor's degree in Chemistry from the Federal University of Pará (UFPA, 2014), and a Master's (2016) and Ph.D. (2018) in Food Science and Technology from the same institution. His doctoral research focused on the extraction of essential oils from different plant matrices using supercritical CO_2 and conventional techniques. He completed a postdoctoral fellowship in Biological Sciences with an emphasis on Tropical Botany at UFRA/MPEG (2023–2025). He has authored 266 academic publications, including 130 articles in international journals, 14 books published internationally by Bentham Science, IntechOpen, De Gruyter, CRC Press/Taylor & Francis, and Springer Nature, 60 international book chapters, and 62 conference abstracts. His expertise includes the extraction of secondary metabolites from medicinal and aromatic plants using supercritical fluid extraction, hydrodistillation, steam distillation, and solvent extraction, as well as classical open-column and thin-layer chromatography. A specialist in natural products chemistry, particularly essential oils and modern analytical identification of secondary metabolites, his research interests encompass food science and technology, pharmacognosy, volatile bioactive molecule prospecting, natural product biotechnology, allelopathy and agricultural applications, and pharmacological and medical studies. These include antibacterial, antiviral, antifungal, antiparasitic, anti-inflammatory, antiproliferative, and antioxidant activities evaluated through in vitro, in vivo, and in silico approaches. Dr. Oliveira has established international collaborations with institutions in Tunisia, China, Belgium, Romania, India, Pakistan, and Italy. He previously served as a permanent professor in the Graduate Program in Biological Sciences (Tropical Botany) at MPEG/UFRA, supervising three Master's students, and as a collaborating professor in Food Science and Technology. He is currently a Visiting Professor in the Graduate Program in Pharmaceutical Sciences at UFPA. He has extensive editorial experience, having reviewed more than 667 manuscripts for about 90 international journals, served as a Bentham Science ambassador and Web of Science mentor, and edited over 400 articles. He holds h-indices of 27 (Web of Science), 30 (Scopus), and 37 (Google Scholar), and was listed among the world's most influential researchers in 2025 (Stanford/Elsevier ranking).

Prof. Dr. Ravendra Kumar is an Assistant Professor in the Department of Chemistry, G.B. Pant University of Agriculture & Technology, Pantnagar, India. He has published over 135 research articles, 22 book chapters, and edited 3 books, including Biorationals and Biopesticides (de Gruyter, 2024) and Natural Pesticides and Allelochemicals (CRC Press, 2024). His research focuses on bioactive natural plant products, including essential oils, terpenes, flavonoids, and alkaloids, as well as nanoformulations targeting insects, nematodes, and fish pathogens. He holds an Indian patent for a portable RP-HPLC device and leads national projects funded by UCB and UCOST. Dr. Kumar has received multiple awards, including the Radhakrishnan Best Teacher-Researcher Award (2025) and Excellence in Research Awards (2023, 2025).

Prof. Dr. Hayet Edziri currently works at the Higher Institute of Biotechnology of Monastir (Tunisia) as an Associate Professor. Her expertise is in natural product research, biomolecules, plant physiology, antimicrobials, agri-food quality, medicinal plants, inflammation, and cell death. She has a Master's degree in Biology and Health and a Ph.D in Biological and Biotechnological Sciences. Since 2012, she has worked as an Associate Professor at the Higher Institute of Biotechnology of Monastir (Tunisia). She has authored many scientific publications, including 60 international journal articles. She is a reviewer and a member of the editorial board of several international journals of scientific papers, with extensive editorial experience and an H-index of 17 in Web of Science, 19 in Scopus, and 23 in Google Scholar.

Prof. Dr. Eloisa Helena de Aguiar Andrade holds a degree in Pharmacy (1980) and a qualification in Biochemistry (1982) from the Federal University of Pará (UFPA). She earned her Master's degree in Natural Products Chemistry (1992) and her Ph.D. in Chemistry (2008), both from UFPA. She is currently a Full Researcher (Senior Researcher II) in the Botany Coordination at the Museu Paraense Emílio Goeldi, where she is responsible for the Adolpho Ducke Multiuser Laboratory (LAD). She also serves as an Associate Professor I at the Faculty of Chemistry at UFPA. In addition, she is a faculty member of the Graduate Programs in Chemistry (UFPA), Biological Sciences - Tropical Botany (UFRA/MPEG), and Biodiversity and Biotechnology (BIONORTE Network). She has coordinated the Pará State branch of the Graduate Program in Biodiversity and Biotechnology (PPG-BIONORTE-PA) during the periods 2016-2020 and 2022-2024. Professor Andrade has authored more than 500 scientific contributions, including journal articles, conference communications, book chapters, and books. Her expertise lies in Natural Products Chemistry, with emphasis on the analysis of volatile and fixed chemical constituents. She has extensive experience in gas chromatography (GC) and gas chromatography-mass spectrometry (GC-MS), particularly in the characterization of derivatized and non-derivatized compounds from natural sources. Her work has significantly contributed to the chemical profiling of Amazonian biodiversity and the development of analytical methodologies for natural products research.

Prof. Dr. Enéas de Andrade Fontes Júnior holds a Bachelor's degree in Pharmacy (2001) with a qualification in Biochemistry (2004) from the Federal University of Pará (UFPA). He earned both his Master's (2004) and Ph.D. (2015) in Neuroscience and Cell Biology from UFPA. He is currently an Associate Professor I at the Institute of Health Sciences at UFPA, working in the School of Pharmacy, where he teaches Pharmacotherapeutic Process and Clinical Pharmacology II. He is a permanent faculty member of the Graduate Programs in Pharmaceutical Sciences (PPGCF, since 2015), Pharmacology and Biochemistry (Farmabio, since 2022), and Pharmaceutical Care (PPGAF, since 2018). He has supervised 22 undergraduate students, 14 Master's students, and one postdoctoral fellow. He is a member of the Brazilian Association of Pharmaceutical Sciences (ABCF) and the Brazilian Society of Medicinal Plants (SBPM). He currently serves as Secretary of the National Forum of Graduate Program Coordinators in Pharmacy (CAPES Area 19) for the 2023–2025 term and as President of the Technical Working Group on Pharmaceutical Education of the Regional Pharmacy Council of Pará (2024–2025). His administrative experience includes serving as Vice Dean (2010–2012) and Dean (2016–2018) of the UFPA School of Pharmacy, President of the UFPA Research Ethics Committee (2017–2021), Vice Coordinator of the Multiprofessional Residency Program in Critical Patient Care (2019–2021), Coordinator of the Graduate Program in Pharmaceutical Sciences (2021–2025), and Head of the Laboratory of Inflammation and Behavior Pharmacology (LAFICO) since 2012. His research focuses on the development of medicines based on biodiversity, particularly in inflammation, neuroinflammation, wound healing, and pain. He conducts studies on the toxicity of natural and synthetic products and evaluates their biological activities related to pain, inflammation, tissue repair, neuroinflammation, and neuropsychiatric disorders.

Contents

Preface

Herbs and spices have accompanied human civilization for thousands of years, serving not only as flavoring agents but also as important components of traditional medicine and cultural practices. Their aromatic properties, preservation potential, and therapeutic uses have made them essential in culinary, medicinal, and agricultural contexts across different societies. In recent decades, scientific interest in herbs and spices has expanded considerably as advances in chemistry, pharmacology, biotechnology, and analytical technologies have revealed the remarkable diversity of bioactive compounds present in these plant materials.

Today, herbs and spices are increasingly recognized as valuable sources of natural products with applications in food science, pharmaceutical development, and biotechnology. Essential oils, phenolic compounds, flavonoids, and other secondary metabolites derived from these plants have demonstrated a wide range of biological activities, including antioxidant, antimicrobial, anti-inflammatory, and therapeutic effects. At the same time, modern technologies are enabling more efficient extraction methods, more precise chemical characterization, and new strategies to enhance the functional value of plant-derived compounds.

This edited volume, *Herbs and Spices - Innovations in Extraction, Chemical Composition, Pharmacology, and Biotechnological Applications*, aims to provide an accessible overview of current scientific advances in these important natural resources. The chapters included in this book cover different aspects of herbs and spices research, ranging from extraction techniques and essential oil composition to pharmacological investigations and biotechnological approaches that enhance the functional properties of plant materials. In addition, modern analytical methodologies are discussed as powerful tools for identifying and characterizing phytochemicals, thereby contributing to a deeper understanding of plant metabolomes.

Another important theme addressed in this volume is the relationship between traditional knowledge and contemporary scientific research. Many herbs and spices have been used for centuries in traditional healing systems, and modern studies are increasingly validating these uses through pharmacological and biochemical investigations. By combining traditional perspectives with modern scientific methods, researchers are discovering new opportunities for developing functional foods, nutraceuticals, and plant-based therapeutic agents.

The chapters also highlight the growing role of biotechnology in transforming herbs and spices into value-added products. Techniques such as microbial fermentation, advanced extraction processes, and high-resolution analytical platforms are expanding the possibilities for improving the bioavailability, stability, and biological effectiveness of plant-derived compounds.

This book is intended for researchers, graduate students, and professionals working in fields such as natural product chemistry, pharmacology, food science, biotechnology, and plant sciences. By presenting diverse scientific perspectives, the volume aims to contribute to a broader understanding of how herbs and spices can support innovation in health, nutrition, and sustainable bio-based technologies. As Academic Editor, I would like to express my sincere appreciation to all contributors and reviewers who generously shared their expertise and helped shape this volume. Their work reflects the growing global interest in exploring the scientific potential of herbs and spices and highlights the importance of interdisciplinary collaboration in natural product research. We hope that this book will serve as a useful reference and encourage further research into the many possibilities offered by these remarkable plant resources.

Prof. Mozaniel Santana de Oliveira, Ph.D.
Postgraduate Program in Pharmaceutical Sciences (PPGCF),
Institute of Health Sciences,
Federal University of Pará,
Belém, Brazil

Prof. Ravendra Kumar, Ph.D.
Department of Chemistry,
C.B.S&.H., G.B.Pant University of Agriculture & Technology,
Uttarakhand, India

Prof. Hayet Edziri, Ph.D.
Laboratory of Transmissible Diseases and
Biologically Active Substances (LR99ES27),
Faculty of Pharmacy,
University of Monastir,
Monastir, Tunisia

Prof. Eloisa Helena de Aguiar Andrade, Ph.D.
Campus de Pesquisa - Museu Paraense Emílio Goeldi,
Belém, Brazil

Prof. Dr. Enéas de Andrade Fontes Júnior, Ph.D.
Postgraduate Program in Pharmaceutical Sciences (PPGCF),
Institute of Health Sciences,
Federal University of Pará,
Belém, Brazil

Chapter 1

Cymbopogon Nardus Essential Oils: Extraction, Chemical Composition and Biological Activities

Caroline da Costa Callandrine, Oberdan Oliveira Ferreira, Marcelo Alves Farias, Fábio José Bonfim Cardoso, Mozaniel Santana de Oliveira, Karyme do Socorro de Souza Vilhena and Eloisa Helena de Aguiar Andrade

Abstract

Essential oils (EOs) have attracted considerable scientific interest due to their ecological role in plants and their broad biological applications. This chapter presents an overview of extraction techniques, chemical composition, and biological properties of EOs, with emphasis on *Cymbopogon nardus* (L.) Rendle. Conventional and modern extraction methods are discussed, highlighting factors that influence oil yield and quality. The genus *Cymbopogon* is recognized for its medicinal relevance and high EO content, particularly in monoterpenes such as citronellal, citronellol, and geraniol, which are responsible for diverse biological activities. The chemical profile of *C. nardus* oil underpins its applications in pharmaceutical, cosmetic, and agricultural industries. Its biological potential includes antioxidant, antimicrobial, and insecticidal properties, which reinforce its value as a natural alternative to synthetic products. The compilation of studies presented in this chapter demonstrates that *C. nardus* EO is a promising candidate for environmentally friendly applications and human health protection.

Keywords: essential oils, *Cymbopogon nardus*, extraction methods, monoterpenes, antioxidant activity, antimicrobial activity, insecticidal potential

1. Introduction

In recent decades, the study of essential oils (EOs) has garnered mounting attention due to their therapeutic properties and industrial applications. These volatile, aromatic compounds are synthesized from various plant parts, including flowers, leaves, seeds, and roots. These substances are the result of secondary metabolism. In contrast to primary metabolites, which are directly involved in plant growth and development, secondary metabolites, such as EOs, have adaptive functions. As demonstrated in the literature, these functions include defense against herbivores and pathogens, as well as the attraction of pollinators [1–4].

EOs are natural, liquid, and highly volatile substances. EOs are produced and stored in specialized structures called secretory cavities, glandular trichomes, and idioblasts. These structures are located in plant organs such as leaves and flowers, and they play a fundamental ecological role for the plant [5, 6]. EOs have a complex chemical composition consisting mainly of low-molecular-weight volatile organic substances and may contain more than 300 different compounds, such as terpenes and phenylpropanoids [7]. The concentrations of these compounds usually depend on the specific characteristics of each plant.

The terpene class constitutes the most substantial and varied category of natural organic compounds, with a total of 80,000 terpenoids identified to date [1, 8]. Terpenes fulfill pivotal biological functions in living organisms, including defense against biotic and abiotic stresses. These compounds are classified as cyclic or acyclic hydrocarbons, with some exhibiting oxygenation [9]. The classification of these compounds is determined by the number of isoprene units (C_5H_8) and the length of the carbon chains [10]. The chemical composition of these substances varies according to the number of isoprene units in their structure. Monoterpenes, with 10 carbon atoms ($C_{10}H_{16}$), and sesquiterpenes, with 15 carbon atoms ($C_{15}H_{24}$), are the main constituents [11, 12].

During the process of obtaining EOs, a variety of factors have the capacity to impact their chemical composition, which, in turn, affects their properties. The variability of these factors, both intrinsic and extrinsic, directly influences the yield and quality of the oil. Among the intrinsic factors, the interaction between species and environmental conditions, such as the time of collection and water stress, stands out. Extrinsic factors, such as the extraction method, precipitation conditions, and postharvest care, have been identified as influential elements in this regard. These factors have the potential to influence not only the composition of EO but also its efficacy [5, 13–15].

The EOs have demonstrated applications in various domains due to their biological activities, including antioxidant, insecticidal, antimicrobial, and antifungal properties [16, 17]. These oils have a variety of applications, including their use as natural food preservatives, in aromatherapy for the treatment of depression and the improvement of sleep quality, and in agriculture as an alternative to synthetic products for mosquito control [18–20].

2. Extraction methods

The production of EOs is a complex process whose efficiency depends on numerous factors, such as the handling of the plant material, increased control and quality of equipment, and improved extraction methods, especially techniques with lower energy, solvent, and time costs [21]. These extraction techniques are classified into two categories: traditional (conventional) and modern [22].

Conventional methods for EO extraction involve the use of water distillation by heating, such as hydrodistillation and steam distillation. Conversely, innovative techniques are predicated on enhancing the extraction process with respect to time criteria for isolating EOs, as well as the yield and quality of these EOs. Among these techniques are supercritical fluid extraction and solvent-free microwave extraction (SFME) [23].

2.1 Hydrodistillation

Hydrodistillation is recognized as one of the most widely used techniques for extracting EOs. In this method, the plant material is immersed in water and brought to a boil, usually at atmospheric pressure. The steam generated carries the volatile compounds and is cooled in a condenser, thus separating the EO from the water. A notable benefit of this approach is that the distillation process takes place at relatively low temperatures, a factor that aids in the preservation of compounds that are particularly sensitive to heat. Despite its prevalence as a technique for distilling botanical material, the method has certain limitations. These include the necessity for greater time and water input, in addition to extended heating of the plant material, which can result in the generation of an undesirable odor in the EO. Moreover, prolonged exposure to boiling water can trigger chemical reactions, such as ester hydrolysis, which involves the transformation of alcohols and acids [24, 25].

2.2 Steam distillation

Steam distillation extraction is a method employed for the extraction of EOs from plant material. This method is regarded as a straightforward and financially feasible process, with the added benefit of not employing toxic solvents that could potentially harm the environment [26]. The duration of distillation can vary from one to 10 hours, depending on factors such as the type of plant, the temperature applied, the exposure time to steam, and the pressure conditions [27]. In this method, the plant material is exposed directly to steam, causing the heat from the steam to break down the plant cells, thereby allowing the release of volatile compounds. Subsequently, the steam is condensed, yielding water and EO. In conclusion, the mixture of EO and water is separated [28].

2.3 Supercritical fluid extraction

This technique is one of the most innovative and sustainable, providing protection to the environment. Furthermore, this method has demonstrated its efficacy in the extraction of EOs from diverse plant matrices. The use of low temperatures during the extraction process ensures the preservation of the active compounds present in EOs, a feature that distinguishes this method from conventional extraction processes. The gas most commonly used in this method is carbon dioxide (CO_2), which is inert and reduces the pressure at the extractor outlet during its elimination [29]. This gas possesses a number of advantageous properties, including its capacity to reach the critical point (with a low critical pressure of 72.9 atm and temperature of 31.2°C) with ease. It is also nonaggressive to plant essence molecules, nonflammable, and available in high purity at a relatively low cost. Additionally, it exhibits a polarity similar to pentane, making it suitable for the extraction of lipophilic compounds [23].

2.4 SFME

This extraction method is regarded as innovative, clean, and environmentally friendly. It is employed for the purpose of obtaining EOs from herbs or aromatic plants. The isolation and concentration of volatile compounds are carried out in a single stage [30]. In addition to being regarded as an environmentally sustainable method, this technique has been shown to yield high levels of EOs while exhibiting selectivity and a reduced extraction time. This is achieved by integrating two techniques: first, microwaves are used to heat the samples, and then atmospheric pressure dry distillation is employed, without the need for solvents [23]. The utilization of these techniques offers several key benefits, including a reduction in energy consumption, a decrease in extraction time, and the elimination of hazardous solvents. At SFME, the chemical constituents of interest are extracted using microwaves that heat and extract the EO from the plant material [31].

In the context of the extraction of the EO from the *Cymbopogon nardus* plant, studies show that the above extraction techniques directly influence the quality of this EO. In addition, there are no statements about the best method used to obtain the volatiles of *C. nardus*. However, it should be noted that extraction methods involving heating can favor the triggering of significant thermal degradations or the hydrolysis of thermosensitive constituents.

3. Genus *Cymbopogon*

The genus *Cymbopogon*, which is part of the Poaceae family, comprises approximately 180 species. These species are native to the tropical and subtropical regions of Asia. Presently, they are cultivated in countries in South and Central America, Africa, and other tropical regions [32].

In Brazil, the genus comprises eight recognized species and 10 synonyms, including *C. caesius*, *C. citratus*, *C. densiflorus*, *C. flexuosus*, *C. iwarancusa*, *C. marginatus*, *C. martini*, *C. natus*, *C. schoenanthus*, and *C. winterianus*. The species of the genus were introduced to Brazil and currently have geographical distributions across the North, Northeast, Midwest, Southeast, and South regions [33].

The species of the genus *Cymbopogon* have been observed to employ a variety of reproductive strategies, including vegetative propagation by means of seeds or rhizomes. However, the process of reproduction by seeds is less efficient due to the rarity of flowering and fruiting in many species. The method of reproduction is contingent upon the species and environmental conditions. The genus's capacity for versatile reproduction has been demonstrated to contribute to its wide distribution and facilitate cultivation [32]. *Cymbopogon* species have been observed to interact with other plants in the environment, engaging in competition for resources such as nutrients, light, and water. Furthermore, they are utilized as livestock feed in agroecological systems [34].

Cymbopogon, a genus of significant medicinal importance, is employed in traditional medicine for the treatment of various ailments. In Brazil, the leaves of *C. marginatus* are utilized in the treatment of epilepsy and anxiety [35]. In East Asia, the leaves of *C. citratus* and *C. flexuosus* have been utilized by traditional peoples for the treatment of digestive problems and fevers [36, 37]. In Egypt, *C. schoenanthus* is utilized in traditional medicine [38]. These traditional applications reflect the empirical knowledge of traditional peoples in different countries, demonstrating the importance of the genus in folk medicine [39].

The genus *Cymbopogon* is distinguished by its capacity to produce substantial amounts of EO in its secondary metabolism, a characteristic that confers upon it significant commercial value. These oils find application in the cosmetic and pharmaceutical industries, as well as in biological control practices [40, 41]. The EOs of the genus are characterized by a high content of terpenes, including hydrocarbons and oxygenated monoterpenes, such as citral, geraniol, myrcene, limonene, and linalool [42–44]. On the other hand, the minority compounds present in the EO, such as β-myrcene, α-terpinolene, 4-terpineol, and other compounds, through synergism, can enhance its application and pharmacological potential.

Finally, it is important to note that the amount and content of volatile compounds present in the EO of the genus *Cymbopogon* during the extraction process are impacted by several factors, both biotic, such as stress caused by fungi and bacteria, and abiotic, such as temperature, soil type, plant nutrition, and sunlight.

4. *Cymbopogon nardus* (L.) Rendle

Cymbopogon nardus (L.) Rendle, commonly known as lemongrass, is a species widely distributed throughout tropical and subtropical regions globally. This species is native to South Asia, particularly India and Sri Lanka, and it has demonstrated a high degree of adaptability to diverse climatic conditions, a factor that has contributed to its successful dispersal and establishment in a variety of environments [45, 46].

C. nardus is cultivated in various regions worldwide, including South America, Africa, India, and China, due to the high commercial interest in its EO, which is produced in the plant's secretory cavities. These specialized structures have been shown to be responsible for both the synthesis and storage of EOs [47, 48].

The species *C. nardus* is notable for its economic potential in industries such as pharmaceuticals, cosmetics, and agriculture. The chemical composition of the oil is widely recognized for the presence of monoterpenes, such as citronellal, geraniol, and citronellol (**Figure 1**). These monoterpenes have been shown to confer antimicrobial, antifungal, antiinflammatory, and insect-repellent properties [2, 47, 49, 50].

Despite the prevalence of monoterpenes in *C. nardus* EO, sesquiterpenes, encompassing both oxygenated and hydrocarbon compounds, are present in trace

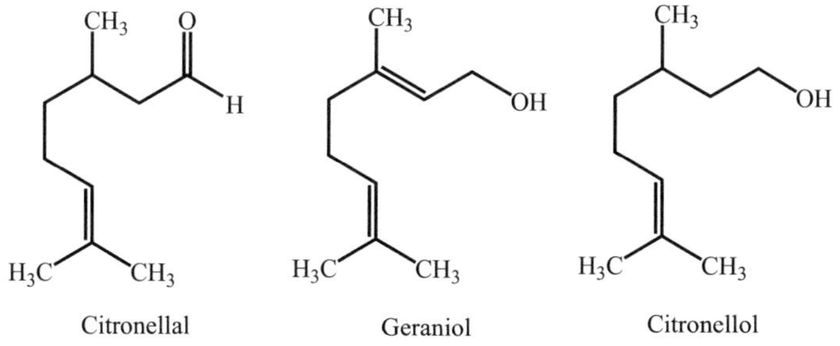

Figure 1.
Two-dimensional structures of citronellal, geraniol, and citronellol.

amounts within the EO of this species. The minimal concentration of these compounds in *C. nardus* oil indicates that their contributions to the oil's overall biological properties are comparatively negligible when compared to monoterpenes [49–52].

The prevalence of oxygenated monoterpenes, predominantly geraniol and citronellal, distinguishes *C. nardus* EO as a promising candidate for applications involving insect repellency, antimicrobial, and antioxidant activities. However, the presence of low concentrations of sesquiterpenes may impose limitations on the utilization of the EO in contexts requiring biological activities associated with these compounds, such as sustained antiinflammatory effects. This chemical profile renders *C. nardus* oil particularly well-suited for formulations aimed at environmental protection and human health.

5. Antioxidant activity

Plants have been identified as a promising source of molecules with antioxidant properties. In this context, the antioxidant activity of EOs can serve as a natural alternative to synthetic antioxidants. EOs are natural, nontoxic, and have fewer side effects, whereas synthetic antioxidants can pose potential risks to human health. Antioxidants are defined as substances with the ability to delay or prevent oxidative damage by neutralizing free radicals. Free radicals are unstable molecules that cause damage to cells and DNA, which can lead to degenerative diseases in the body [53, 54].

There are several methods employed to assess the antioxidant capacity of a substance. The $ABTS^+$ assay is a particularly noteworthy method that is used to assess the antioxidant capacity of different substances based on the neutralization of the stable radical cation ($ABTS^+$). The $ABTS^+$ radical is distinguished by its blue-green hue and demonstrates efficacy for both hydrophilic and lipophilic compounds. The $DPPH^•$ test is predicated on the neutralization of the 2,2-diphenyl-1-picrylhydrazyl radical ($DPPH^•$) by electron donation, where the color change indicates antioxidant activity [55].

The antioxidant activity of *C. nardus* EO was examined using two radical-based assays: $DPPH^•$ and 2,2'-azino-bis (3-ethylbenzothiazoline-6-sulfonic acid) ($ABTS^+$). The findings of the study demonstrated that the EO manifested antioxidant properties by impeding the action of $DPPH^•$ radicals and the $ABTS^+$ radical cation. In the $DPPH^•$ assay, the results indicated 102.19 ± 4.2 μg extract/μg compared to 0.11 ± 0.04 μg for gallic acid, which was used as a control. In the $ABTS^+$ assay, EO demonstrated activity of 0.009 ± 0.0004 μM TE/g, in comparison to 2.66 ± 0.31 μM TE/g for the control group. The authors attribute the antioxidant potential to the presence of geraniol, with a concentration of 28.40% in the oil. Conversely, the EO exhibited reduced activity in comparison to gallic acid, which was utilized as a standard reference [50].

The antioxidant potential of *C. nardus* oil was investigated using the $DPPH^•$ method. The study's findings indicated that the oil exhibited a high antioxidant capacity, with values reaching 93.86 μg TEAC/mL. This outcome may be associated with the bioactivity of compounds such as geraniol and limonene [56].

The difference observed in the results of the oxidizing activity of these EOs is related to the different methods and the mediums in which they react, for example, ethanol for the DPPH assay and phosphate-buffered saline for the ABTS assay. In addition, there is another factor, such as reaction time (5 minutes for the ABTS assay and 30 minutes for the DPPH assay).

6. Biological potential

6.1 Insecticide and toxicity

The term "insecticidal activity" refers to the capacity of certain compounds, including EOs, to repel, control, or eliminate insects. This activity offers effective solutions for pest control in agriculture and can help combat insect resistance to synthetic insecticides. These compounds are generally nontoxic to mammals and act as fumigants and repellents. Consequently, EOs have emerged as a promising alternative for the development of safer insecticides. A notable example is *Cymbopogon nardus* oil, which contains active compounds such as geraniol, citronellal, and limonene. These compounds have the capacity to interfere with the neural pathways of insects through various mechanisms of action, making it a highly effective insecticide [2, 57–60].

Boonyuan et al. [61] evaluated the insecticidal potential of the EO from *C. nardus* leaves against the *Aedes aegypti* mosquito. The researchers found a median lethal concentration (LC_{50}) of 4.02%, indicating low toxicity of the oil. This result suggests that *C. nardus* is less effective at killing mosquitoes compared to other EOs tested, such as *C. citratus* oil.

In a related study, the larvicidal activity of *C. nardus* EO was examined against *A. aegypti* larvae, demonstrating effective larvicidal action with an LC_{50} of 75.85 mg/mL. The chemical composition of the EO was characterized by citronellal (21.8%), citronellol (18.1%), and geraniol (11.3%) [62]. Studies by Manimaran et al. [63] and Kurniasih et al. [64] demonstrated that *C. nardus* EOs exhibited toxicity against *A. aegypti* larvae at 47.21 mg/L and 35.133 mg/L, respectively, over a 24-hour period. Consequently, the utilization of EO as an alternative to pesticides is a viable proposition.

The insecticidal potential of *C. nardus* EO was investigated against larvae of two genotypes of *Anopheles gambiae*, a mosquito that hosts and transmits malaria [65]. The EO exhibited high larvicidal activity, with LC_{50} values of 36.83 mg/L and 97.33 mg/L in two distinct strains. According to the authors, citronellal (37.87%) and nerol (19.88%) may be responsible for the larvicidal activity of the oil [65].

The insecticidal and repellent activity of the EO was evaluated against two insect pests, *Oryzaephilus surinamensis* (L.) and *Sitophilus zeamais* Motsch. The repellent activity values after four hours of exposure were determined to be $LC_{50} = 0.03\ \mu L/cm^2$ (*O. surinamensis*) and $LC_{50} = 0.04\ \mu L/cm^2$ (*S. zeamais*). For fumigant activity, the EO of *C. nardus* ranged from 46.9 to >604 μL/L. Although the EO of *C. nardus* exhibited low toxicity in relation to fumigant activity against the two insect pests, it was observed that it has repellent action [49]. Loko et al. [66] evaluated the repellent activity of *C. nardus* EO against *Dinoderus porcellus*, and the results demonstrated that even low concentrations of the EO can act as a repellent.

The activity of EO against *Artemia salina* larvae was examined, revealing that the oil exhibited no toxicity, with LC_{50} values greater than 1,000 μg/mL. The oil contained concentrations of citronellal (27.34%), geraniol (23.21%), and geranial

(13.37%) [67]. In contrast, in the study by Cansian et al. [68], citronella EO was evaluated against *A. aegypti* and *A. salina* larvae. The activity values were LC_{50} = 111.84 μg/mL and LC_{50} = 5.29 μg/mL, respectively, with concentrations of citronellal (38.98%), geraniol (20.56%), and citronellol (17.54%).

6.2 Antimicrobial

The antimicrobial activity of EOs has been highlighted in several studies due to their potential to combat the action of microorganisms. These oils contain complex bioactive compounds that possess antifungal and antibacterial properties. The antibacterial efficacy of essential oils can vary depending on the plant from which they are derived and the type of target bacteria, being influenced by the chemical structure of the compounds present, their functional groups, and synergistic interactions. Research indicates that the primary antimicrobial action occurs through the inhibition of cell growth or the death of bacterial cells. Furthermore, these oils have been the focus of increased research as potential alternatives to combat bacterial resistance, demonstrating their value in the development of novel treatments [54, 69, 70].

The antifungal activity of *C. nardus* EO was evaluated against *Candida albicans* in vitro and in vivo, with the minimum inhibitory concentration (MIC) determined using the microdilution technique. The oil was found to contain citronellal (27.34%), geraniol (23.21%), and geranial (13.37%), which are known to exhibit antifungal activity. The MIC values for the standard strain (CAATCC) and the clinical strain (CAV) were determined to be 500 μg/mL and > 500 μg/mL, respectively. The oil's efficacy was demonstrated in an in vivo study [67]. Prado et al. [71] evaluated the antifungal activity of an EO from *C. nardus* with a high citronellal (44.44%) and citronellol (16.97%) content against *C. albicans*, and the substance demonstrated efficacy against two strains (planktonic and biofilm). The authors reported that the EO contains constituents that aid in the antifungal effect and that it is an alternative for the control of fungal infections.

The EO of *Cymbopogon nardus*, collected in Piraquara, Brazil, was evaluated for its antifungal potential *in vitro* against mycotoxin-producing strains of *Aspergillus niger*, *A. nomius*, *A. flavus*, and *Fusarium graminearum*. The EO, which is abundant in geraniol (35.52%), citronellal (24.39%), and citronellol (10.73%), demonstrated antimicrobial activity against all fungi tested through contact and volatilization methods, with MIC values of 0.75%, 0.75%, 0.5%, and 0.25% [72].

A number of studies have demonstrated the antibacterial properties of *C. nardus* EO in combating pathogens. The antibacterial activity of *C. nardus* EO was evaluated against *Escherichia coli* strains, with MIC determined using the microdilution technique. The oil exhibited no substantial effects against the 22 *E. coli* strains, with MIC of ≥6,400 μg/mL [73]. As demonstrated in studies conducted by Millezi et al. [74], the oil exhibited antibacterial properties against *Staphylococcus aureus* and *E. coli* bacteria. This effect was attributed to the presence of elevated concentrations of citronellal.

In other studies, the EO was evaluated for its antibacterial action against 10 species of *Pseudomonas* isolated from freshwater fish. Utilizing disc diffusion and MIC methodologies, the oil exhibited favorable properties for impeding the proliferation of these species, which are frequently linked with the deterioration of foodstuffs such as fish [56].

7. Conclusions

This chapter has therefore presented a significant compilation of information about EOs, also demonstrating the main techniques for obtaining them. The chapter has considered both conventional methods, such as hydrodistillation and steam distillation, and modern methods, such as supercritical fluid extraction and microwave extraction. Furthermore, the text underscores the significance of the genus *Cymbopogon* and the species *Cymbopogon nardus* in the production of EO, a matrix distinguished by the presence of oxygenated monoterpenes, including citronellal, geraniol, and citronellol. Moreover, it elucidates numerous pertinent studies that document the antioxidant potential of *C. nardus* EO, along with its antimicrobial and insecticidal properties.

Author details

Caroline da Costa Callandrine[1], Oberdan Oliveira Ferreira, Marcelo Alves Farias[2], Fábio José Bonfim Cardoso[1], Mozaniel Santana de Oliveira[3]*, Karyme do Socorro de Souza Vilhena[1] and Eloisa Helena de Aguiar Andrade[1]

1 Faculty of Chemistry, Institute of Exact and Natural Sciences, Federal University of Pará, Belém, Brazil

2 Federal University of Pará, Belém, Brazil

3 Laboratory of Pharmacology of Inflammation and Behavior, Institute of Health Sciences, Federal University of Pará, Belém, Brazil

*Address all correspondence to: mozaniel.oliveira@yahoo.com.br

References

[1] Nwokeji PA, et al. The Chemistry of Natural Product: Plant Secondary Metabolites. International Journal of Technology Enhancements and Emerging Engineering Research. 2016;**4**(8):1. https://www.researchgate.net/publication/307858399

[2] Kaur H, Bhardwaj U, Kaur R. Cymbopogon Nardus Essential Oil: A Comprehensive Review on Its Chemistry and Bioactivity. Journal of Essential Oil Research. 2021;**33**:205–220. DOI: 10.1080/10412905.2021.1871976

[3] Gurav TP, Dholakia BB, Giri AP. A glance at the chemodiversity of Ocimum species: Trends, implications, and strategies for the quality and yield improvement of essential oil. Springer Science and Business Media B.V. 2022. DOI: 10.1007/s11101-021-09767-z

[4] Hegazy MM, Mekky RH, Ibrahim AE, Abouelela ME, Kedra TA, Al-Harrasi A. Essential Oils: The Science of Extraction and Its Implications for Composition and Biological Activity—A Review. Food Analytical Methods. 2025;**18**:1483–1513. DOI: 10.1007/s12161-025-02808-9

[5] Dhifi W, Bellili S, Jazi S, Bahloul N, Mnif W. Essential Oils' Chemical characterization and investigation of some biological Activities: A Critical Review. Medicines. 2016;**3**(n°4):25. DOI: 10.3390/medicines3040025

[6] Sharifi-Rad J, et al. Biological activities of essential oils: From plant chemoecology to traditional healing systems. MDPI AG. 2017. DOI: 10.3390/molecules22010070

[7] de Sousa DP, de Assis Oliveira F, Arcanjo DDR, da Fonsêca DV, Duarte ABS, de Oliveira Barbosa C, Ong TP, Brocksom TJ. Essential Oils: Chemistry and Pharmacological Activities—Part II. Biomedicines. 2024;**12**:1185. DOI: 10.3390/biomedicines1206118

[8] Christianson DW. Structural and chemical biology of terpenoid cyclases. American Chemical Society. 2017. DOI: 10.1021/acs.chemrev.7b00287

[9] Barros DB, et al. Antifungal activity of terpenes isolated from the Brazilian Caatinga: A review. Brazilian Journal of Biology. 2023;**83**. DOI: 10.1590/1519-6984.270966

[10] Koshima CC, Umeda TK, Nakamoto KT, Venâncio LL, Aracava KK, Rodrigues CEC. (Liquid + Liquid) equilibrium for systems composed of clove and allspice essential oil compounds and hydrous ethanol at T = 298.2K. Journal of Chemical Thermodynamics. 2016;**95**:54–62. DOI: 10.1016/j.jct.2015.10.023

[11] Mozaniel O, Sousa-Filho AP, De Oliveira MS, Da Silva Souza Filho AP, editors. Terpenoids: Recent Advances in Extraction, Biochemistry and Biotechnology. vol. **1**. Sharjah, U.A.E: Benthamscience; 2021. 400 p. DOI: 10.2 174/97816810896451220101

[12] Guimarães AC, et al. Antibacterial activity of terpenes and terpenoids present in essential oils. Molecules. 2019;**24**(n° 13):2471. DOI: 10.3390/molecules24132471

[13] Zeng Z, Zhang S, Wang H, Piao X. Essential oil and aromatic plants as feed additives in non-ruminant nutrition: A

review. BioMed Central. 2015. DOI: 10.1186/s40104-015-0004-5

[14] Boaro CSF, Vieira MAR, Campos FG, Ferreira G, De-la-cruz-chacón I, Marques MOM. Factors influencing the production and chemical composition of essential oils in Aromatic Plants from Brazil. In: Malik S, editor. Essential Oil Research: Trends in Biosynthesis, Analytics, Industrial Applications and Biotechnological Production. Cham: Springer International Publishing; 2019. p. 19–47. DOI: 10.1007/978-3-030-16546-8_2

[15] Ribeiro FP, Santana de Oliveira M, de Oliveira Feitosa A, Santana Barbosa Marinho P, Moacir do Rosario Marinho A, de Aguiar Andrade EH, Favacho Ribeiro A. Chemical Composition and Antibacterial Activity of the Lippia Origanoides Kunth Essential Oil from the Carajás National Forest, Brazil. Evidence-Based Complement. Alternative Medicine. 2021:1–8. DOI: 10.1155/2021/9930336

[16] Pandey VK, et al. Exploiting the bioactive properties of essential oils and their potential applications in food industry. The Korean Society of Food Science and Technology. 2023. DOI: 10.1007/s10068-023-01287-0

[17] Zhao Q, Zhu L, Wang S, Gao Y, Jin F. Molecular mechanism of the anti-inflammatory effects of plant essential oils: A systematic review. Elsevier Ireland. 2023. DOI: 10.1016/j.jep.2022.115829

[18] Bolouri P, et al. Applications of Essential Oils and Plant Extracts in Different Industries. MDPI. 2022. DOI: 10.3390/molecules27248999

[19] Faheem F, et al. Uncovering the industrial potentials of lemongrass essential oil as a Food Preservative: A Review. MDPI. 2022. DOI: 10.3390/antiox11040720

[20] Yin XJ, Lin GP, Wu XY, Huang R, Xu CJ, Yao MY. Effects of lavender essential oil inhalation aromatherapy on depression and sleep quality in stroke patients: A single-blind randomized controlled trial. Complementary Therapies in Clinical Practice. 2024;**55**:101828. DOI: 10.1016/j.ctcp.2024.101828

[21] Kusuma HS, Mahfud M. Kinetic studies on extraction of essential oil from sandalwood (Santalum album) by microwave air-hydrodistillation method. Alexandria Engineering Journal. 2018;**57**(n° 2):1163–1172. DOI: 10.1016/j.aej.2017.02.007

[22] Baldino L, Scognamiglio M, Reverchon E. Supercritical fluid technologies applied to the extraction of compounds of industrial interest from Cannabis sativa L. and to their pharmaceutical formulations: A review. Journal of Supercritical Fluids. 2020;**165**:104960. 104960. DOI: 10.1016/j.supflu.2020.104960

[23] Aziz ZAA, et al. Essential Oils: Extraction techniques, pharmaceutical and therapeutic potential-A review. Current Drug Metabolism. 2018;**19** (n° 13):1100–1110. DOI: 10.2174/1389200219666180723144850

[24] Marques TM, de Souza FM, Espinheira MJCL. Revisão Bibliografica Sobre Os Diferentes Métodos De Extração De Óleo Essencial, em Princípios Físico Químicos em Farmácia. Atena Editora. 2019;147–154. DOI: 10.22533/at.ed.41319051113

[25] Katekar VP, Rao AB, Sardeshpande VR. Review of the rose essential oil extraction by hydrodistillation: An investigation for the optimum operating condition for maximum yield. Sustainable Chemistry and Pharmacy. 2022;**29**:100783. DOI: 10.1016/j.scp.2022.100783

[26] Valentim JA, Soares EC. Extração de Óleos Essenciais por Arraste a Vapor: Um Kit Experimental para o Ensino de Química. Química Nova Na Escola. 2018;297–301. DOI: 10.21577/0104-8899.20160131

[27] Danlami JM, Arsad A, Zaini MAA, Sulaiman H. A comparative study of various oil extraction techniques from plants. Reviews in Chemical Engineering. 2014;**30**(n° 6). DOI: 10.1515/revce-2013-0038

[28] de Oliveira MS. Essential Oils. Cham: Springer International Publishing; 2022. DOI: 10.1007/978-3-030-99476-1

[29] Silva SG, et al. Supercritical CO2 extraction to obtain Lippia thymoides Mart. & Schauer (Verbenaceae) essential oil rich in thymol and evaluation of its antimicrobial activity. Journal of Supercritical Fluids. 2021;**168**:105064. 105064. DOI: 10.1016/j.supflu.2020.105064

[30] Khalili G, Mazloomifar A, Larijani K, Tehrani MS, Azar PA. Solvent-free microwave extraction of essential oils from Thymus vulgaris L. and Melissa officinalis L. Industrial Crops and Products. 2018;**119**:214–217. DOI: 10.1016/j.indcrop.2018.04.021

[31] Rana GMM, et al. Assessment of the essential oil extracted from Citrus aurantifolia leaves using solvent-free microwave extraction technique. Food Chemistry Advances. 2025;**6**:100913. DOI: 10.1016/j.focha.2025.100913

[32] Lawal OA, Ogundajo AL, Avoseh NO, Ogunwande IA. Cymbopogon citratus. In: Victor Kuete, editor. Medicinal Spices and Vegetables from Africa. vol. **1**. New York: Elsevier; 2017. p. 397–423. DOI: 10.1016/B978-0-12-809286-6.00018-2

[33] Cymbopogon in Flora e Funga do Brasil. Jardim Botânico do Rio de Janeiro. Available at: https://floradobrasil.jbrj.gov.br/FB615041&#x003E [Accessed: 2025-March-26]

[34] Wanapat M, Cherdthong A, Pakdee P, Wanapat S. Manipulation of rumen ecology by dietary lemongrass (Cymbopogon citratus Stapf.) powder supplementation1. Journal of Animal Science. 2008;**86**(n° 12):3497–3503. DOI: 10.2527/jas.2008-0885

[35] Rbara LSLB, et al. Volatile constituents and behavioral change induced by Cymbopogon winterianus leaf essential oil in rodents. African Journal of Biotechnology. 2011;**10**(n° 42):8312–8319. DOI: 10.5897/AJB10.509

[36] Jeong M-R, Park PB, Kim D-H, Jang Y-S, Jeong HS, Choi S-H. Essential Oil prepared from cymbopogon citrates exerted an antimicrobial activity against plant pathogenic and medical Microorganisms. Mycobiology. 2009;**37** (n° 1):48–52. DOI: 10.4489/MYCO.2009.37.1.048

[37] Desai MA, Parikh J. Microwave Assisted Extraction of Essential Oil From Cymbopogon Flexuosus (Steud.) Wats: A Parametric and Comparative Study. Separation Science and Technology. 2012;**47**(n° 13):1963–1970. DOI: 10.1080/01496395.2012.659785

[38] El-Bakry AA. Regeneration from embryogenic callus and suspension cultures of the wild medicinal plant Cymbopogon schoenanthus. African Journal of Biotechnology. 2012;**11** (n° 43). DOI: 10.5897/AJB12.331

[39] Avoseh O, Oyedeji O, Rungqu P, Nkeh-Chungag B, Oyedeji A. Cymbopogon Species; Ethnopharmacology, phytochemistry and the pharmacological importance. Molecules. 2015;**20**(n° 5):7438–7453. DOI: 10.3390/molecules20057438

[40] Shah G, Shri R, Panchal V, Sharma N, Singh B, Mann A. Scientific basis for the therapeutic use of Cymbopogon citratus, stapf (Lemon grass). Journal of Advanced Pharmaceutical Technology & Research. 2011;**2**(n°1):3. DOI: 10.4103/2231-4040.79796

[41] Mukarram M, et al. Lemongrass Essential Oil Components with Antimicrobial and Anticancer Activities. Antioxidants. 2021;**11** (n° 1):20. DOI: 10.3390/antiox11010020

[42] Ju J, Xie Y, Guo Y, Cheng Y, Qian H, Yao W. Application of edible coating with essential oil in food preservation. Critical Reviews in Food Science and Nutrition. 2019;**59** (n° 15):2467–2480. DOI: 10.1080/10408398.2018.1456402

[43] Li M, Liu B, Bernigaud C, Fischer K, Guillot J, Fang F. Lemongrass (Cymbopogon citratus) oil: A promising miticidal and ovicidal agent against Sarcoptes scabiei. PLoS Neglected Tropical Diseases. 2020;**14**(n° 4): e0008225. DOI: 10.1371/journal.pntd.0008225

[44] Wani AR, Yadav K, Khursheed A, Rather MA. An updated and comprehensive review of the antiviral potential of essential oils and their chemical constituents with special focus on their mechanism of action against various influenza and coronaviruses. Microbial Pathogenesis. 2021;**152**:104620. 104620. DOI: 10.1016/j.micpath.2020.104620

[45] Hamzah MH, Man HC, Abidin ZZ, Jamaludin EH. Citronella oil extraction. 2014

[46] Brito LCF, et al. Analysis of the chemical composition, antifungal activity and larvicidal action against Aedes aegypti larvae of the Essential Oil Cymbopogon nardus. Research, Society and Development. 2021;**10**(n° 13): e543101321452. DOI: 10.33448/rsd-v10i13.21452

[47] Nakahara K, Alzoreky NS, Yoshihashi T, Nguyen HTT, Trakoontivakorn G. Chemical Composition and Antifungal Activity of Essential Oil from Cymbopogon nardus (Citronella Grass). Japan Agricultural Research Quarterly: JARQ. 2013;**37** (n° 4):249–252. DOI: 10.6090/jarq.37.249

[48] Pavela R, Benelli G. Essential oils as Ecofriendly Biopesticides? Challenges and Constraints. Trends in Plant Science. 2016;**21**(n° 12):1000–1007. DOI: 10.1016/j.tplants.2016.10.005

[49] Hernandez-Lambraño R, Pajaro-Castro N, Caballero-Gallardo K, Stashenko E, Olivero-Verbel J. Essential oils from plants of the genus Cymbopogon as natural insecticides to control stored product pests. Journal of Stored Products Research. 2015;**62**:81–83. DOI: 10.1016/j.jspr.2015.04.004

[50] Bayala B, et al. Chemical composition, antioxidant,

anti-inflammatory and antiproliferative activities of the essential oil of Cymbopogon nardus, a plant used in traditional medicine. Biomolecular Concepts. 2020;**11**(n° 1):86–96. DOI: 10.1515/bmc-2020-0007

[51] Chizzola R. Regular Monoterpenes and Sesquiterpenes (Essential Oils). In: Ramawat KG, Jean-Michel Merillon, editors. Natural Products. vol. **1**. Berlin, Heidelberg: Springer Berlin Heidelberg; 2013. p. 2973–3008. DOI: 10.1007/978-3-642-22144-6_130

[52] Kumoro AC, Wardhani DH, Retnowati DS, Haryani K. A brief review on the characteristics, extraction and potential industrial applications of citronella grass (Cymbopogon nardus) and lemongrass (Cymbopogon citratus) essential oils. IOP Conference Series: Materials Science and Engineering. 2021;**1053**(n° 1):012118. 012118. DOI: 10.1088/1757-899X/1053/1/012118

[53] Shahidi F, Zhong Y. Measurement of antioxidant activity. Elsevier. 2015. DOI: 10.1016/j.jff.2015.01.047

[54] Basavegowda N, Baek KH. Synergistic antioxidant and antibacterial advantages of essential oils for food packaging applications. MDPI AG. 2021. DOI: 10.3390/biom11091267

[55] Munteanu IG, Apetrei C. Analytical methods used in determining antioxidant activity: A review. MDPI AG. 2021. DOI: 10.3390/ijms22073380

[56] Kačániová M, et al. The antioxidant and antimicrobial activity of essential oils against Pseudomonas spp. isolated from fish. Saudi Pharmaceutical Journal. 2017;**25**(n° 8):1108–1116. DOI: 10.1016/j.jsps.2017.07.005

[57] Ma S, Jia R, Guo M, Qin K, Zhang L. Insecticidal activity of essential oil from Cephalotaxus sinensis and its main components against various agricultural pests. Industrial Crops and Products. 2020;**150**:112403. DOI: 10.1016/J.INDCROP.2020.112403

[58] Olivo CJ, et al. Óleo de citronela no controle do carrapato de bovinos. Ciência Rural. 2008;**38**(n° 2):406–410. DOI: 10.1590/S0103-84782008000200018

[59] Ong M, Chomistek N, Dayment H, Goerzen W, Baines D. Insecticidal activity of plant powders against the Parasitoid, Pteromalus venustus, and its Host, the Alfalfa Leafcutting Bee. Insects. 2020;**11**(n° 6):359. DOI: 10.3390/insects11060359

[60] Zimmermann RC, et al. Insecticide activity and toxicity of essential oils against two stored-product insects. Crop Protection. 2021;**144**:105575. DOI: 10.1016/J.CROPRO.2021.105575

[61] Boonyuan W, Grieco JP, Bangs MJ, Prabaripai A, Tantakom S, Chareonviriyaphap T. Excito-repellency of essential oils against an Aedes aegypti (L.) field population in Thailand. Journal of Vector Ecology. 2014;**39**(n° 1):112–122. DOI: 10.1111/j.1948-7134.2014.12077.x

[62] Ríos N, Stashenko EE, Duque JE. Evaluation of the insecticidal activity of essential oils and their mixtures against Aedes aegypti (Diptera: Culicidae). Revista Brasileira de Entomologia. 2017;**61**(n°4):307–311. DOI: 10.1016/j.rbe.2017.08.005

[63] Manimaran A, Cruz MMJJ, Muthu C, Vincent S, Ignacimuthu S. Larvicidal and knockdown effects of some essential oils against Culex

quinquefasciatus Say, Aedes aegypti (L.) and Anopheles stephensi (Liston). Advances in Bioscience and Biotechnology. 2012;**03**(n° 07):855–862. DOI: 10.4236/abb.2012.37106

[64] Kurniasih N, Nuryadin W, Harahap MN, Supriadin A, Kinasih I. Toxicity of essential oils from Orange (Citrus sinesis L. Obbeck) and lemongrass (Cymbopogon nardus L. Rendle) on Aedes aegypti a vector of Dengue Hemorrhagic Fever (DHF). Journal of Physics: Conference Series. 2021;**1869**(n° 1):12015. 012015. DOI: 10.1088/1742-6596/1869/1/012015

[65] Ahouansou AC, Fagla SRM, Tokoudagba JM, Toukourou H, Badou YK, Gbaguidi FA. Chemical composition and larvicidal activity of the essential oil of Cymbopogon nardus (L.) Rendle on Anopheles gambiae. International Journal of Biological and Chemical Sciences. 2019;**13**(n° 3): 1861–1869. DOI: 10.4314/ijbcs.v13i3.53

[66] Loko YLE, et al. Bioactivity of essential oils of Cymbopogon citratus (DC) Stapf and Cymbopogon nardus (L.) W. Watson from Benin against Dinoderus porcellus Lesne (Coleoptera: Bostrichidae) infesting yam chips. International Journal of Tropical Insect Science. 2021;**41**(n° 1):511–524. DOI: 10.1007/s42690-020-00235-3

[67] de Toledo LG, et al. Improved in vitro and in vivo anti-candida albicans activity of Cymbopogon Nardus essential oil by its incorporation into a microemulsion system. International Journal of Nanomedicine. 2020;**15**:10481–10497

[68] Cansian RL, et al. Toxicity and larvicidal activity on Aedes aegypti of citronella essential oil submitted to enzymatic esterification. Brazilian Journal of Biology. 2023;**83**. DOI: 10.1590/1519-6984.244647

[69] Chouhan S, Sharma K, Guleria S. Antimicrobial activity of some Essential Oils—Present Status and Future Perspectives. Medicines. 2017;**4**(n° 3):58. DOI: 10.3390/medicines4030058

[70] de Sousa DP, et al. Essential Oils: Chemistry and Pharmacological Activities. Biomolecules. 2023;**13**(7): 1144. DOI: 10.3390/biom13071144

[71] Prado GM, et al. Antifungal, molecular docking and cytotoxic effect of the essential oil of Cymbopogon citratus (DC) Stapf. and Cymbopogon nardus (L.) Rendle against Candida albicans. Anais da Academia Brasileira de Ciências. 2024;**96**(n° 3). DOI: 10.1590/0001-3765202420230309

[72] Zimmermann RC, et al. Antifungal activity of essential oils and their combinations against storage fungi. Environmental Science and Pollution Research. 2023;**30**(n° 16):48559–48570. DOI: 10.1007/s11356-023-25772-5

[73] Lara VM, Carregaro AB, Santurio DF, de Sá MF, Santurio JM, Alves SH. Antimicrobial Susceptibility of Escherichia coli Strains Isolated from Alouatta spp. Feces to Essential Oils. Evidence-Based Complementary and Alternative Medicine. 2016;**2016**(n° 1). DOI: 10.1155/2016/1643762

[74] Millezi AF, Baptista NN, Caixeta DS, Rossoni DF, Cardoso MG, Piccoli RH. Caracterização química e atividade antibacteriana de óleos essenciais de plantas condimentares e medicinais contra Staphylococcus aureus e Escherichia coli. Revista Brasileira de Plantas Medicinais. 2014;**16**(n° 1):18–24. DOI: 10.1590/S1516-05722014000100003

Chapter 2

Fermentation of Herbs and Spices: A Biotechnological Strategy to Enhance Functional Value

Raquel Gómez-Pliego, Judith Espinosa-Raya, Hulme Ríos-Guerra and Jair Alejandro Temis-Cortina

Abstract

This chapter explores microbial fermentation as a biotechnological strategy to enhance the functional and nutraceutical properties of herbs and spices. Although these vegetable materials are naturally rich in bioactive compounds such as polyphenols, terpenes, and organosulfur compounds, their efficacy is often constrained by poor bioavailability, instability, or metabolic inactivation in the human body. Controlled microbial fermentation offers a promising approach to overcome these limitations by promoting the biotransformation of complex phytochemicals into more bioactive, soluble, or absorbable forms, thereby enhancing their physiological impact. This chapter includes a comprehensive overview of traditional, advanced, and assisted separation techniques used to isolate bioactive constituents, as well as hybrid and specialized methodologies. It also addresses key variables that influence the phytochemical profiles of herbs and spices prior to fermentation. Illustrative case studies involving turmeric, ginger, garlic, clove, and oregano are discussed to demonstrate how fermentation can reduce antinutritional factors and increase the release or synthesis of health-promoting metabolites. Microbial species such as lactic acid bacteria, Saccharomyces spp., and Bacillus spp. are analyzed in terms of their functional roles and process requirements. This chapter concludes with an evaluation of current applications in functional foods, fermented beverages, and dietary supplements that must be considered for their industrial implementation. Special emphasis is placed on the role of fermentation in expanding access to bioavailable, evidence-based botanical products with potential public health benefits.

Keywords: fermented herbs, spices, biotechnology, microbial fermentation, health-promoting metabolites, andantioxidant activity

1. Introduction

Herbs and spices have been used since ancient times not only to enhance the flavor, aroma, and color of food but also for their medicinal properties [1]. In recent decades, interest in these plants has grown substantially in both scientific and clinical fields due to their richness in bioactive compounds with recognized antioxidant, anti-inflammatory, antimicrobial, and other health-promoting effects [2, 3]. Despite their potential, many of these phytochemicals face significant challenges, such as low bioavailability, poor stability under physiological conditions, and limited in vivo efficacy [4]. Moreover, although numerous studies describe their chemical composition and biological effects, there are still important gaps regarding their mechanisms of action in humans, the influence of food matrices on their bioactivity, and the lack of standardized clinical trials that validate their long-term safety and efficacy. These compounds are responsible for a broad spectrum of physiological effects that contribute to the therapeutic value of herbs and spices. They have been associated with the prevention and mitigation of chronic diseases, modulation of oxidative stress, and enhancement of immune responses. However, their effectiveness often depends on their chemical structure, concentration, and interactions with other dietary components – aspects that remain insufficiently explored in translational and clinical research.

A schematic representation of the health-related benefits of herbs and spices is shown in **Figure 1**. This diagram not only illustrates their functional roles but also highlights the most common limitations that hinder their application in clinical and nutritional settings, particularly due to compound instability and low bioavailability.

The figure outlines five key dimensions of herb and spice functionality: (1) culinary purposes, including their ability to enhance flavor, color, and aroma in food; (2) food preservation, particularly through extending the shelf life of pickled or fermented products; (3) medical purposes, as traditionally used in herbal medicine

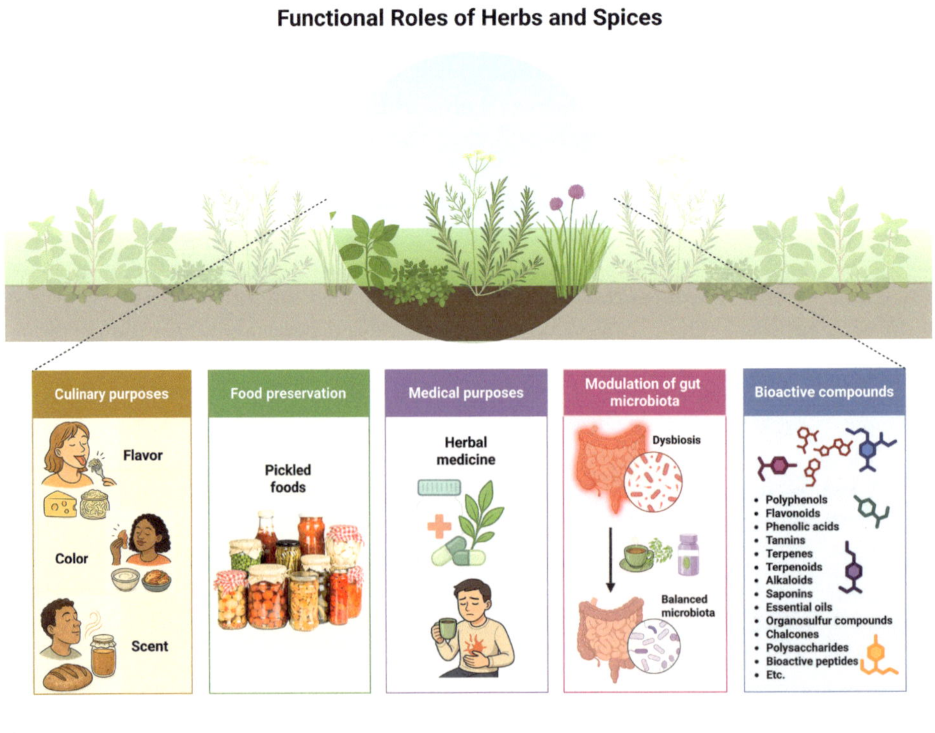

Figure 1.
Multifunctional roles and bioactive compounds of herbs and spices.

to treat various ailments; (4) modulation of gut microbiota, helping correct dysbiosis and support a balanced intestinal ecosystem; and (5) the presence of diverse bioactive compounds, such as polyphenols, flavonoids, phenolic acids, tannins, terpenes, alkaloids, saponins, essential oils, organosulfur compounds, chalcones, polysaccharides, and bioactive peptides.

Together, these elements emphasize the multifunctionality of herbs and spices, linking their culinary value to significant nutraceutical and therapeutic potential. However, their clinical use remains limited unless strategies are developed to improve the stability and bioavailability of these bioactive compounds.

Microbial fermentation emerges as an innovative biotechnological strategy to overcome the limitations associated with the traditional use of herbs and spices. Through controlled fermentation processes, the compounds present in these natural products can be transformed to improve their solubility, bioavailability, and biological activity. Nevertheless, more evidence is needed to clarify which microbial strains and fermentation conditions are most effective, and how these processes influence the safety, sensory profile, and regulatory acceptance of the final products. This chapter aims to examine the mechanisms by which microbial fermentation enhances the functional and nutraceutical properties of herbs and spices. Additionally, it will address clean technology-based methods for extraction and purification, review relevant scientific studies, and assess their current applications and prospects in the food, pharmaceutical, and nutraceutical industries.

2. Bioactive composition of herbs and spices

Herbs and spices are abundant in a wide range of bioactive compounds, primarily polyphenols (including flavonoids and phenolic acids), terpenes, alkaloids, sulfur-containing compounds, tannins, and essential oils. These secondary metabolites are largely responsible for their beneficial biological effects. For example, curcumin in turmeric, gingerol in ginger, eugenol in clove, and allicin in garlic have demonstrated strong antioxidant and anti-inflammatory activities [5–7].

A summary of the main groups of bioactive compounds commonly identified in herbs and spices is illustrated in **Figure 2**. This includes flavonoids, terpenoids, alkaloids, tannins, phenolic acids, resistant starches, bitter compounds, sulfur-containing compounds, and essential oils, among others. These metabolite groups are closely associated with a wide range of biological activities, such as antioxidant, anti-inflammatory, antimicrobial, and metabolic regulatory effects. Their structural diversity underpins the therapeutic potential and functional properties of botanical products applied in food, pharmaceutical, and nutraceutical formulations, highlighting their relevance for health-promoting interventions [8].

However, many of these compounds exhibit low oral bioavailability due to factors such as poor water solubility, first-pass hepatic metabolism, or transformation by the intestinal microbiota. Additionally, some of these compounds may be present in conjugated or "inactive" forms that require release or modification to exert their biological effects [9].

Table 1 below presents a summary of the main bioactive compounds found in commonly used herbs and spices [10]. It includes their most studied functional effects, such as antioxidant, anti-inflammatory, and antimicrobial activities, as well

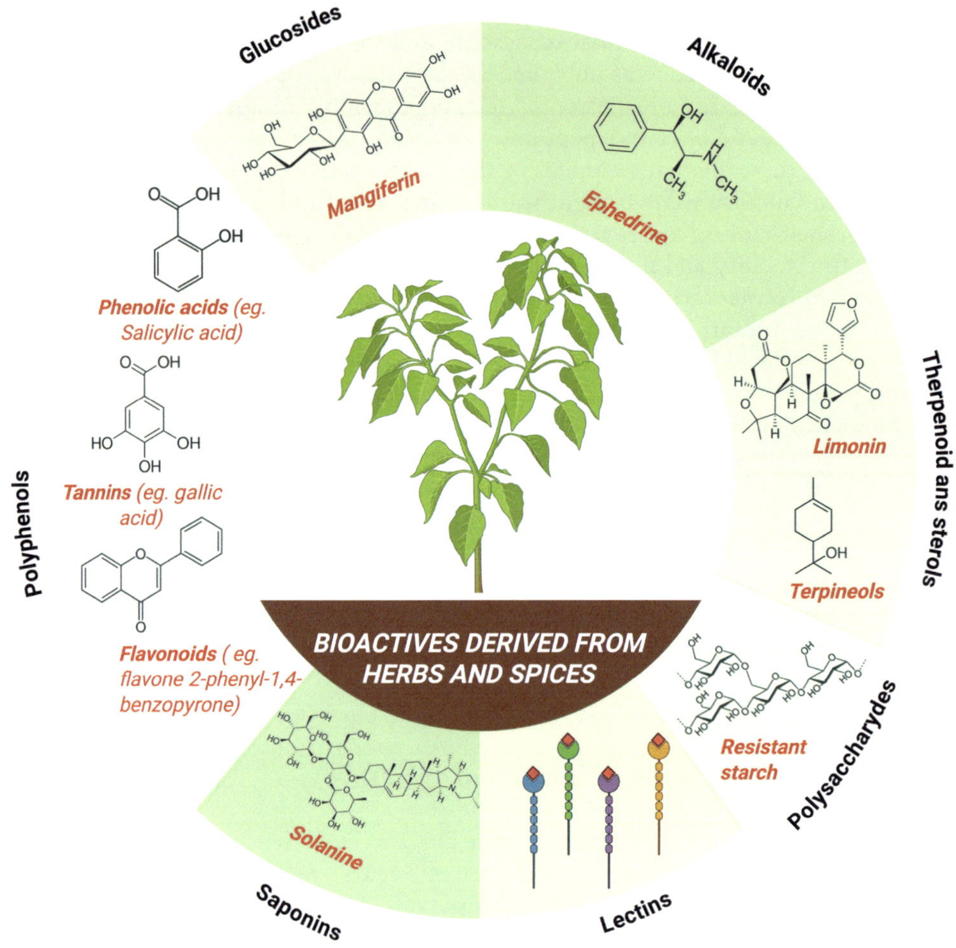

Figure 2.
Bioactive compounds in herbs and spices: classification of the main groups, structural diversity, and functional properties of bioactive metabolites commonly identified in herbs and spices.

Herb/spice	Main compound	Functional property	Natural limitation
Turmeric	Curcumin	Antioxidant, anti-inflammatory	Low solubility and bioavailability
Ginger	Gingerol	Antiemetic, antioxidant	Thermal instability
Garlic	Allicin	Antimicrobial, hypocholesterolemic	Instability during processing
Clove	Eugenol	Antioxidant, analgesic	Volatility and easy oxidation
Rosemary	Carnosic acid, carnosol	Anti-inflammatory, neuroprotective	Rapid oxidation

Table 1.
Functional properties and natural limitations of major bioactive compounds from selected herbs and spices.

as typical limitations related to their natural form, including low solubility, instability, or poor bioavailability under physiological conditions.

3. Fundamentals of microbial fermentation

Microbial fermentation is a metabolic process in which microorganisms, such as bacteria, yeasts, and filamentous fungi, convert organic substrates into secondary metabolites under anaerobic or microaerophilic conditions. Traditionally, this process has been used for food preservation, flavor enhancement, and improved digestibility. More recently, it has gained recognition for its ability to modify the biochemical profile of fermented ingredients and increase the bioavailability of phytochemicals.

The most commonly employed microorganisms include the following:

Lactic acid bacteria (LAB) (e.g., Lactobacillus, Leuconostoc, Pediococcus, Lactococcus)

They are highly efficient in producing organic acids (mainly lactic acid), bacteriocins, and bioactive peptides. Their enzymatic systems hydrolyze glycosidic bonds, liberating phenolic aglycones with higher bioactivity. However, their metabolic spectrum is relatively narrow, limiting the transformation of highly complex compounds.

Yeasts (e.g., Saccharomyces cerevisiae, Kluyveromyces spp.)

They play a central role in aroma and flavor development through the production of ethanol, higher alcohols, and esters. Yeasts also secrete extracellular enzymes capable of modifying polyphenols, but their contribution to enhancing mineral bioavailability or reducing antinutrients is less pronounced compared to bacteria.

Filamentous fungi (*e.g., Aspergillus oryzae, Rhizopus, Penicillium* spp.)

These organisms exhibit a broad enzymatic repertoire (amylases, proteases, tannases, cellulases) that allows for the degradation of complex macromolecules such as tannins, saponins, or polymerized phenolics. As a result, they can release novel metabolites with higher bioactivity. Nevertheless, their use requires stricter process control due to potential risks of contamination and secondary metabolite production (e.g., mycotoxin).

During fermentation, enzymes produced by these microorganisms break down complex structures such as polysaccharides, glycosides, and tannins, thereby releasing bioactive compounds or converting them into more soluble, bioavailable, and biologically active forms [11]. In addition, new metabolites (including organic acids, bioactive peptides, and exopolysaccharides) are generated, further contributing to the functional potential of the final product.

4. Fermentation applied to herbs and spices

As previously mentioned, herbs and spices have been used since ancient times for their culinary, medicinal, and preservative properties. Today, their application has expanded to include biotechnological and functional uses, particularly in the development of fermented foods. The incorporation of these botanical ingredients not only enhances the sensory profile of the final product's flavor, aroma, and color but also improves its microbiological stability and nutraceutical potential [12, 13].

To ensure that these benefits are preserved throughout fermentation, it is essential to understand and control the processing and extraction methods used for the bioactive compounds in herbs and spices, as well as the variables that influence their chemical composition and biological activity. These aspects are summarized in **Figure 3**.

4.1 Extraction methods

To fully harness the functional potential of herbs and spices in fermented products, it is essential to apply appropriate extraction methods that allow the isolation of their bioactive compounds. These procedures directly influence the availability, stability, and efficacy of metabolites such as polyphenols, flavonoids, and essential oils. Extraction can be broadly classified into traditional methods and advanced/assisted techniques, each with particular advantages and limitations [14–20].

4.1.1 Traditional methods

Maceration consists of soaking plant material in solvents (water, ethanol, hexane, etc.) for 2–3 days at room temperature with agitation, allowing passive

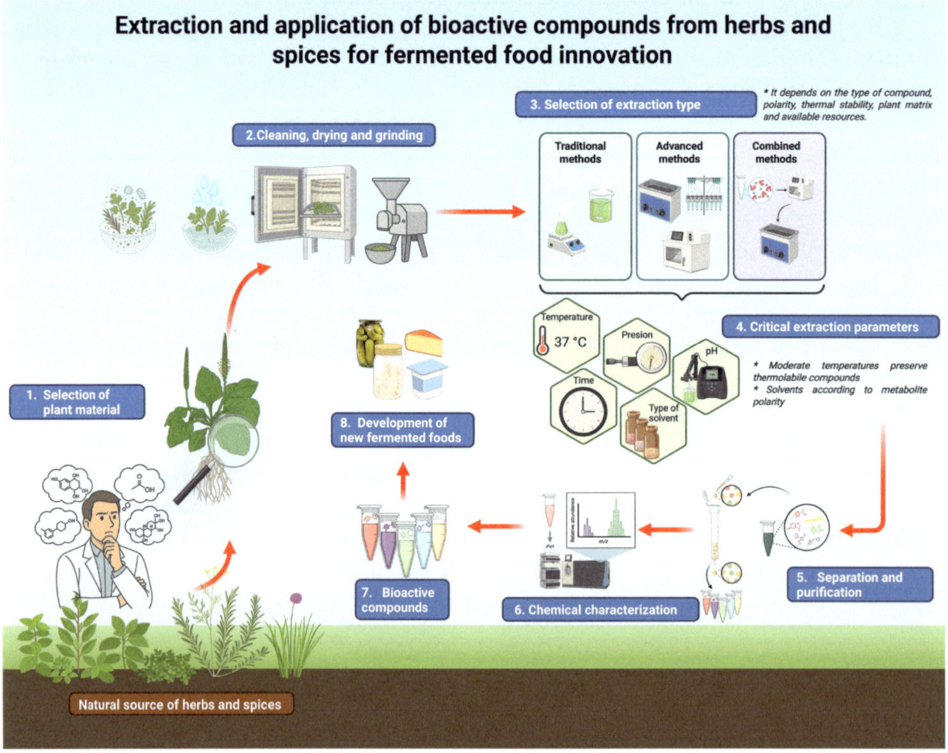

Figure 3.
Extraction and application of bioactive compounds from herbs and spices for fermented food innovation.

diffusion of phytochemicals. It is simple and inexpensive but slow and low-yielding [14, 15].

Digestion resembles maceration but incorporates mild heating (35–50°C for 30 minutes to 24 hours), which improves the solubility of certain polyphenols without significant degradation, provided they tolerate moderate temperatures [14].

Infusion involves steeping plant material in hot or cold water for short periods, which is common in medicinal teas. Extracts may be concentrated afterward by vacuum evaporation [15].

Decoction requires boiling plant material in water (15–60 minutes), which is especially effective for thermostable compounds in rigid tissues such as roots or barks [15].

Percolation passes hot solvent through plant material at a controlled rate (5–7 drops/min), yielding better extraction than maceration, though it consumes more solvent and requires technical skill [14].

Soxhlet extraction employs continuous solvent recirculation in a closed-loop system, which is efficient for thermostable metabolites (phenols, lipids) but unsuitable for thermolabile compounds and is energy-intensive [16].

Hydrodistillation, widely used for essential oils, employs water or steam distillation followed by solvent trapping (pentane/diethyl ether 1:1 v/v). It works well with matrices such as clove and ginger but requires long processing times and high energy [17].

4.1.2 Advanced and assisted methods

Ultrasound-assisted extraction (UAE) applies ultrasonic cavitation (20–2000 kHz, ~ 30 min) to disrupt cell walls and release compounds such as phenolic acids and flavonoids, often using 70% methanol. It is rapid, eco-friendly, and energy-efficient, though free radical formation may occur [14].

Microwave-assisted extraction (MAE) heats polar solvents (ethanol/water) using microwave radiation, causing cell rupture. Typically performed with 50% ethanol at 180–1000 W for 2–25 minutes, it is effective for phenols, flavonoids, and essential oils [17, 18].

Pressurized liquid extraction uses high temperatures (50–200 °C) and pressures (35–200 bar) below supercritical conditions to enhance the recovery of phenols, oils, and alkaloids with low solvent consumption, and is considered a green, efficient method [19].

Supercritical fluid extraction exploits supercritical CO_2 ($T > 31°C$; $P > 7.38$ MPa), which is ideal for essential oils, terpenoids, and alkaloids. It is selective and residue-free but requires costly equipment [14, 18].

Pulsed electric field applies short, high-voltage pulses to induce electroporation, improving the release of sugars, polyphenols, and carotenoids while conserving energy. It is widely applied in juices, oils, and wines to increase yield, clarity, and sensory quality [20].

Enzyme-assisted extraction (EAE) uses hydrolytic enzymes (cellulases, pectinases) to break down plant cell walls, releasing flavonoids and lipids. Although eco-friendly, it requires careful control of pH, temperature, and enzyme dosage, and can be expensive [14, 16, 18].

Other emerging techniques include turbodistillation, which applies shear forces to release essential oils with reduced thermal damage; countercurrent extraction, improving diffusion with opposite flows of solvent and material

(~25 °C); solid-phase extraction, fast and sensitive for trace analytes but limited by sorbent capacity; and high-voltage-assisted extraction, where electric fields disrupt cells with minimal solvent use [14].

4.1.3 Combined and specialized methods

To further optimize yield and selectivity, combined techniques are being developed. Ultrasound–microwave assisted extraction (UMAE) integrates ultrasonic cavitation and microwave heating, achieving high extraction rates in short times with low solvent volumes, suitable for saponins and alkaloids [18].

Enzyme–ultrasound–microwave-assisted extraction (EUMAE) combines EAE, UAE, and MAE, maximizing the recovery of delicate compounds from complex matrices, though it remains under investigation [18].

Liquid–liquid extraction partitions analytes between immiscible solvents (e.g., water/ethyl acetate, water/hexane), which is useful for volatiles and antioxidants, often as pretreatment [14].

Solid-phase microextraction employs coated fibers to adsorb analytes directly from samples. It is solvent-free, automated, and highly sensitive, especially for trace compounds [14].

Instant controlled pressure drop (DIC) uses moderate pressure (~0.35 MPa), followed by sudden decompression, expanding plant tissues to enhance the release of anthocyanins and polyphenols. It improves yields and texture, though specialized equipment is required [6, 14].

4.2 Variables affecting the phytochemical profile of herbs and spices

The phytochemical profile of herbs and spices, mainly composed of secondary metabolites such as flavonoids, terpenoids, and phenolic compounds, is highly sensitive to various factors. These factors determine both the qualitative and quantitative composition of bioactive compounds, directly influencing their functionality and bioavailability. Key variables include the following:

Plant part used: The distribution of secondary metabolites varies significantly across roots, stems, leaves, flowers, and fruits. For example, in Curcuma longa, curcumin is concentrated in the rhizomes, while in Ocimum basilicum, flavonoids and essential oils are abundant in the leaves [21]. Tissue localization directly affects extract bioactivity, making organ selection critical.

Cultivation conditions: Soil type, sunlight exposure, altitude, and water availability all influence the biosynthesis of phenolic compounds and terpenoids. In Mentha piperita, UV radiation and water stress have been shown to increase menthol and rosmarinic acid production [22]. Organic farming also tends to enhance phytochemical content due to higher biotic stress.

Postharvest drying and processing: Drying methods affect the preservation of volatile and thermolabile compounds. Drying at temperatures above 60 °C can degrade flavonoids and essential oils, like eugenol in Syzygium aromaticum (clove), reducing their bioactivity [23]. Shade drying or lyophilization is preferred for preserving phytochemical integrity.

Grinding: Particle size affects surface area and extraction efficiency. However, excessive grinding can cause oxidation due to oxygen exposure, decreasing the stability of compounds such as anthocyanins and catechins [24].

Solvent type: Solvent polarity plays a key role in the selective solubilization of metabolites. Glycosylated flavonoids and phenolic acids are best extracted with hydroalcoholic solutions, while lipophilic terpenoids require nonpolar solvents such as hexane [25]. Binary mixtures (e.g., ethanol/water) are commonly used to optimize phytochemical recovery.

pH of the medium: pH affects the stability of phenolic compounds. Phenolic acids, such as ferulic acid, are more stable in slightly acidic media (pH 4 to 6), while alkaline conditions can lead to nonenzymatic oxidation [26].

Extraction temperature and time: These parameters influence diffusion kinetics and compound stability. Moderate temperatures (40 to 60 °C) promote efficient extraction without damaging sensitive compounds. Extending extraction beyond saturation can lead to degradation or polymerization of tannins [27].

Understanding these variables is a key to designing optimal strategies for maximizing the functionality of herbal extracts, particularly in fermentation processes where metabolite availability impacts microbial interaction and the biotransformation of bioactive compounds.

An integrated diagram is presented in **Figure 3**, summarizing the sequential steps for processing herbs and spices aimed at the extraction and application of bioactive compounds in fermented foods. The workflow begins with plant selection, cleaning, and pretreatment, followed by the choice of conventional or advanced extraction techniques, chemical characterization, and subsequent purification steps to ensure the quality and integrity of the metabolites. Critical control parameters are emphasized throughout the process to preserve bioactive compound stability and maximize their functional properties. The final stage involves the incorporation of these purified compounds into fermented products to develop novel functional foods with enhanced nutritional and health-promoting benefits.

4.3 Fermentation of herbs and spices with LAB

Lactic acid bacteria are widely used in the food industry due to their ability to ferment sugars and produce compounds such as lactic acid, bioactive peptides, and bacteriocins. In the context of herbs and spices, their enzymatic activity enables the hydrolysis of conjugated phenolic compounds, releasing more active and bioavailable forms.

For example, the fermentation of turmeric with *Lactobacillus plantarum* has been shown to increase the content of free curcuminoids and enhance antioxidant activity [28]. Similarly, ginger fermented with *Lactobacillus brevis* exhibited greater in vitro anti-inflammatory activity than the crude extract [29].

4.4 Fermentation of herbs and spices with yeasts and fungi

Yeast, such as Saccharomyces cerevisiae, and filamentous fungi, such as *Aspergillus oryzae,* can produce extracellular enzymes that alter the chemical structure of plant-derived compounds. For instance, black garlic is obtained through a controlled thermal fermentation process that increases the concentration of phenolic compounds and reduces allicin content, thereby improving its stability and antioxidant activity [30, 31].

During the fermentation of ginseng with *Aspergillus* spp., the transformation of ginsenosides into minor forms with higher biological activity has been observed, including antitumor and adaptogenic effects [31].

4.5 Mixed fermentation or application in functional beverages

The incorporation of herbs and spices into fermented matrices, such as kombucha, kefir, or tepache, has yielded promising results. These combinations enable synergistic interactions between fermentative microbiota and botanical compounds. For example, kombucha fermented with ginger or turmeric has shown increased levels of glucuronic acid, total polyphenols, and antioxidant capacity compared to the base beverage [32].

In addition, these beverages enhance sensory acceptance and offer functional benefits that exceed the sum of their individual components.

These studies highlight the potential of fermentation to enhance the nutraceutical value of herbs and spices, positioning them as high-value functional ingredients.

Table 2 presents selected examples highlighting the quantitative effects of microbial fermentation on herbs and spices.

These quantitative examples demonstrate how controlled microbial fermentation can significantly enhance the functional quality of botanicals. For instance, fermentation of turmeric increased curcumin levels by approximately 10%, while fermentation of *Moringa oleifera* leaves dramatically reduced antinutrient phytates by nearly 67%, simultaneously improving protein digestibility and mineral availability. A liquid fermentation of *Moringa* also reduced tannins and increased antioxidant capacity and total phenolic content. These results underscore the biochemical potential of fermentation to simultaneously attenuate antinutrients and amplify bioactive compounds in herbs and spices.

5. Reduction of antinutritional factors and enhancement of bioavailability

One of the most notable benefits of microbial fermentation is its capacity to reduce or eliminate antinutritional factors present in herbs and spices. This has a direct impact on improving their bioavailability and functional potential. Antinutritional factors are compounds that interfere with the digestion, absorption, or utilization of essential nutrients and, in some cases, may cause adverse effects if consumed in high concentrations [36].

Plant/matrix	Microorganism	Effect (percent change)	Reference
Turmeric (Curcuma longa)	Lactobacillus fermentum	Curcumin content increased by 9.76%	[33]
Moringa oleifera leaves (old leaves)	Lactobacillus plantarum A6	Phytate content reduced by 66.9%	[34]
Moringa oleifera (liquid fermentation)	Bacillus subtilis SIX-15	Tannin reduced by ~ 17%, DPPH scavenging ↑ 18.8%, total phenols ↑ 35.8%.	[35]

Table 2.
Functional and quantitative impacts of microbial fermentation on bioactive compounds from herbs and spices.

Some of the most relevant antinutritional compounds include the following:

Phytates: Chelate minerals such as calcium, iron, zinc, and magnesium, thereby reducing their intestinal absorption.

Tannins: Form complexes with proteins and minerals, decreasing their bioavailability.

Oxalates: Bind to calcium to form insoluble crystals, which may contribute to kidney stone formation.

Saponins: May damage the intestinal mucosa and impair the absorption of lipids and fat-soluble vitamins.

Phenolic glycosides: Often present in conjugated forms, which limit their absorption in the gastrointestinal tract.

5.1 Mechanisms of reduction through fermentation

During fermentation, microorganisms produce a variety of extracellular enzymes that act on these compounds:

Phytases: hydrolyze phytic acid into inositols and phosphates, facilitating the release of bound minerals.

Tannases: degrade both hydrolyzable and condensed tannins, reducing their capacity to chelate nutrients.

Oxalate decarboxylases: convert oxalates into more soluble and less toxic compounds.

Glycosidases: hydrolyze glycosidic bonds, releasing aglycones, which are more readily absorbed.

These transformations not only eliminate the adverse effects of antinutrients but also generate secondary metabolites with beneficial properties, such as free flavonoids or phenolic acids, which exhibit enhanced bioactivity [37].

5.2 Examples of fermented herbs and spices

Moringa (Moringa oleifera): Fermentation with *Lactobacillus plantarum* significantly reduces phytate and tannin levels while enhancing the bioavailability of iron and calcium. These effects have been confirmed in both in vitro models and animal studies [38].

Basil and oregano: Studies have demonstrated that fermentation with LAB transforms phenolic glycosides into free aglycones, such as luteolin and apigenin – compounds with high antioxidant capacity and improved bioavailability [4].

Rosemary: Rosemary contains polymerized phenolic acids with poor bioaccessibility. Fermentation with fungi, such as *Rhizopus oryzae,* depolymerizes these compounds, generating monomeric derivatives with better intestinal absorption and enhanced anti-inflammatory activity [39].

Fenugreek (Trigonella foenum-graecum): Fermentation reduces the content of steroidal saponins responsible for intestinal irritation, improving the bioactive compound profile and digestive tolerability [40].

5.3 Impact on bioavailability

Improvement in bioavailability not only implies greater absorption but also better biodistribution and systemic biotransformation. Fermentation-derived metabolites have been documented to:

Cross intestinal membranes more effectively (like flavonoid aglycones).

Interact positively with the gut microbiota to promote the production of short-chain fatty acids.

Avoid antagonistic interactions with other nutrients present in the food matrix.

Furthermore, in animal and human bioavailability studies, a significant increase in the plasma concentration of bioactive compounds has been observed after the consumption of previously fermented herbs and spices compared to their nonfermented counterparts [41].

5.4 Nutritional and therapeutic implications

Reducing antinutritional factors through fermentation has important implications for vulnerable populations, such as children, the elderly, and individuals with mineral deficiencies. It is also a strategic tool in the development of functional foods targeted at patients with inflammatory bowel diseases, metabolic syndrome, or micronutrient deficiencies.

Moreover, fermentation contributes to the creation of safer and more effective matrices for the formulation of nutraceutical supplements by eliminating potential adverse effects associated with the consumption of raw or poorly processed plant materials.

In summary, the fermentation of herbs and spices represents an effective biotechnological strategy to enhance their nutritional, functional, and therapeutic value. Through the action of specific microorganisms, it is possible to convert antinutritional compounds into bioactive metabolites with higher bioavailability, reduce gastrointestinal side effects, and promote the generation of peptides and polyphenols with antioxidant, anti-inflammatory, and metabolically protective properties. These transformations also contribute to the positive modulation of the gut–microbiota–brain axis and the development of fermented products with probiotic properties and clinically relevant effects [42]. As illustrated in **Figure 4**, microbial fermentation of herbs and spices through pretreatment and controlled processes enhances the bioavailability of bioactive metabolites while reducing antinutritional factors. Key benefits include improved nutrient transformation (e.g., increased antioxidants, B vitamins, GABA), probiotic potential with gastrointestinal and immunomodulatory effects, functional microbial activities such as gut microbiota modulation and metabolic health improvement, and the production of bioactive compounds with antihypertensive, antidiabetic, anti-inflammatory, and detoxifying properties.

Clinical validation gap

Although fermentation has shown promising results in reducing antinutritional factors and improving the bioavailability of bioactive compounds in vitro and in animal studies, clinical validation in human populations remains limited. Most available evidence is preclinical, and further randomized controlled trials are required to confirm the nutritional and therapeutic implications of these findings under real physiological and dietary conditions.

6. Current and potential applications

The use of microbial fermentation to valorize herbs and spices has opened new opportunities across multiple sectors. The main current and emerging areas of application include functional foods, dietary supplements, natural cosmetics, and

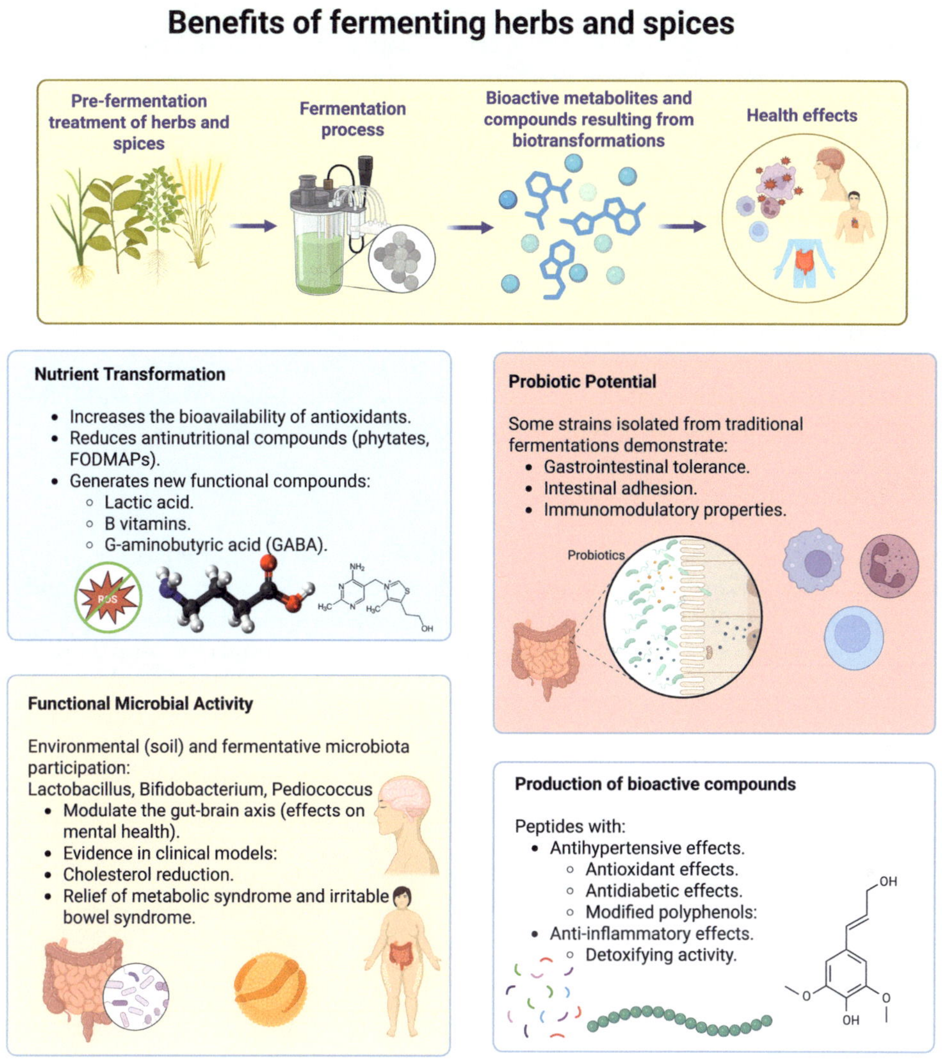

Figure 4.
Main benefits of fermenting herbs and spices, from pretreatment steps to their impact on human health.

pharmaceutical products. Owing to its ability to enhance the bioavailability of active compounds, eliminate antinutritional factors, and generate bioactive metabolites, fermentation of botanicals has gained momentum in markets that demand natural, safe, and scientifically validated solutions [43].

6.1 Food industry

In the food sector, the fermentation of herbs and spices has enabled the development of innovative functional products, such as:

Fermented condiments: These improve digestibility, offer complex flavor profiles, and exhibit enhanced antioxidant properties. Examples include sauces made with black garlic, fermented turmeric, or fermented ginger.

Fermented beverages: Products such as kombucha, kefir, and tepache, enriched with fermented spice infusions (e.g., turmeric, cinnamon, clove, holy basil), combine probiotic effects with additional functional properties.

Functional bakery and confectionery: Fermented extracts are used as fat replacers or natural preservatives, contributing to health-promoting baked goods.

Fortified vegan foods: Fermentation of botanical mixtures can impart umami flavor, reduce antinutritional factors, and enhance the bioavailability of plant-based proteins.

These applications also support the development of clean-label foods, characterized by recognizable ingredients and the absence of synthetic additives – attributes aligned with current consumer preferences for natural products.

6.2 Nutraceutical supplements

In the nutraceutical field, fermentation has been used to improve the bioavailability and efficacy of bioactive compounds, leading to standardized and encapsulated formulations containing highly bioavailable metabolites:

Fermented turmeric capsules, enriched with free curcuminoids and enhanced for oral absorption.

Fermented ginger extracts, exhibit improved anti-inflammatory and digestive activity.

Fermented ginseng extracts, with higher levels of active ginsenosides.

Symbiotic formulations, combining botanically derived prebiotics and probiotic strains.

These fermented matrices have been associated with improved thermal stability, better pharmacokinetic profiles, and fewer gastrointestinal side effects [44].

6.3 Natural cosmetics

In the cosmetics industry, fermented herb and spice extracts have amplified the benefits of natural skincare products. Their antioxidant, antimicrobial, and anti-inflammatory properties are particularly valued in functional and dermocosmetic formulations [45]. Examples include the following:

Creams and tonics with fermented extracts of rosemary, ginseng, licorice, or aloe vera.

Serums containing black garlic or fermented turmeric to combat skin photoaging.

Shampoos and scalp lotions with fermented extracts that support scalp microbiota balance.

Fermentation also improves dermal penetration of active compounds and reduces the irritancy associated with some raw plant extracts [46].

6.4 Pharmaceutical and therapeutic applications

Beyond nutraceutical and cosmetic uses, emerging pharmaceutical and therapeutic applications are being explored, such as:

Gastroprotective formulations, using fermented extracts of basil or mint for gastrointestinal disorders.

Delivery systems for the targeted release of active compounds derived from turmeric, ginger, or rosemary.

Immunomodulatory adjuvants, based on fermented spices, have the potential to modulate pro-inflammatory cytokines (e.g., IL-6, TNF-α).

Metabolic adjuncts for individuals with metabolic syndrome, obesity, or gut dysbiosis.

These products are expected to gain traction in the "fermented functional botanicals" category as clinical evidence expands and regulatory frameworks evolve [47].

6.5 Toward sustainable and functional markets

In addition to functional advantages, fermentation offers significant sustainability benefits. It allows the valorization of plant by-products such as peels, stems, and leaves, enabling the production of high-value compounds with low environmental impact. This approach also promotes circular economies and fair-trade practices, particularly within communities engaged in the cultivation of medicinal plants [48].

Such strategies are especially relevant within the global shift toward more sustainable and resilient food systems, in alignment with the United Nations' Sustainable Development Goals.

6.6 Industrial feasibility and scalability

The industrial application of fermented herbal and spice extracts requires standardized fermentation protocols, cost-effective downstream processing, and regulatory compliance to ensure product safety and consistency. Recent studies highlight that scalable fermentation technologies, including solid-state and submerged bioreactors, enable the controlled production of bioactive metabolites with reproducible quality at the pilot and industrial levels [49, 50]. Moreover, integrating green extraction techniques, such as supercritical CO_2 and enzyme-assisted methods, with fermentation steps enhances process efficiency while reducing energy consumption and solvent use, thus supporting sustainable manufacturing practices [51]. Nevertheless, challenges remain in terms of large-scale validation, shelf-life studies, and alignment with food safety regulations before widespread commercialization can be achieved.

7. Future perspectives

Regulatory challenges and standardization needs

The field of herb and spice fermentation is rapidly expanding, offering new opportunities for the development of sustainable and functional products. Advances in targeted fermentation using specific microbial strains capable of modifying distinct metabolites, together with omics-based approaches such as metabolomics, transcriptomics, and proteomics, are providing deeper insights into biochemical pathways and enabling process optimization. Emerging technologies, including solid-state and semi-solid fermentation, enhance the preservation of sensory qualities and stability of final products, while symbiotic systems combining probiotics and prebiotics derived from fermented botanicals are attracting attention for their potential health benefits. Despite these promising developments, key challenges remain regarding clinical validation, regulatory harmonization, and industrial implementation. At present, most fermented botanical products lack standardized definitions for safety, efficacy, and quality control, resulting in regulatory discrepancies across countries and regions [50, 52]. Establishing clear international guidelines for microbial strain selection, toxicological evaluation, and health claim substantiation, as well as aligning these innovations with existing food safety frameworks such as those from EFSA, FDA, and Codex Alimentarius, will be essential to build consumer confidence, ensure safety, and facilitate large-scale commercialization.

8. Conclusions

Microbial fermentation has emerged as a powerful biotechnological strategy to enhance the functional and nutraceutical properties of herbs and spices. Through the metabolic activity of selected microorganisms, bioactive compounds can be transformed into more bioavailable and biologically active forms, while antinutritional factors are reduced or eliminated. These processes not only preserve but often improve the sensory and health-related qualities of botanical ingredients, supporting their incorporation into functional foods, dietary supplements, and natural health products.

Despite the promising results reported in numerous in vitro and preclinical studies, the translation of these findings into clinically validated interventions remains limited. The lack of standardized fermentation protocols, uniform safety assessments, and harmonized regulatory frameworks across countries creates barriers to reproducibility, consumer trust, and industrial implementation.

Addressing these challenges requires multidisciplinary efforts focused on three essential pillars:

- Clinical validation through well-designed human trials to substantiate health claims;

- Technological standardization to ensure reproducibility, safety, and quality control at an industrial scale; and

- Regulatory harmonization to align innovations with international food safety frameworks such as EFSA, FDA, and Codex Alimentarius.

Only by integrating scientific evidence, technological advances, and regulatory compliance will it be possible to unlock the full potential of microbial fermentation for the development of next-generation functional foods and nutraceuticals with proven efficacy, safety, and global market acceptance.

Acknowledgments

The authors thank Ramón Maruri-Gómez for the language revision. J. Espinosa-Raya acknowledges the Instituto Politécnico Nacional and COFAA-SIP/IPN.

Conflict of interest

The authors declare that they have no conflicts of interest.

Notes/thanks/other declarations

All authors have reviewed and approved the published version of the article.

Author details

Raquel Gómez-Pliego[1*], Judith Espinosa-Raya[2], Hulme Ríos-Guerra[3] and Jair Alejandro Temis-Cortina[1,2]

1 Departamento de Ciencias Biológicas, Sección de Ciencias de la Salud Humana, Laboratorio de Microbiología Industrial, Facultad de Estudios Superiores Cuautitlán, Campo 1, Universidad Nacional Autónoma de México, Cuautitlán Izcalli, México

2 Laboratorio Multidisciplinario de Ciencias Biomédicas, Escuela Superior de Medicina, Instituto Politécnico Nacional, Ciudad de México, México

3 Departamento de Ciencias Químicas, Sección de Química Orgánica, Laboratorio de Química Orgánica Multicomponente, Facultad de Estudios Superiores Cuautitlán, Campo 1, Universidad Nacional Autónoma de México, Cuautitlán Izcalli, México

*Address all correspondence to: pliegoraquel@comunidad.unam.mx

References

[1] Singhal RS, Kulkarni PR. Herbs and spices. Food Authenticity and Traceability, Elsevier. 2003;386–414. DOI: 10.1533/9781855737181.2.386

[2] Filannino P, Di Cagno R, Gobbetti M. Metabolic and functional paths of lactic acid bacteria in plant foods: Get out of the labyrinth. Current Opinion in Biotechnology. 2018;**49**:64–72. DOI: 10.1016/j.copbio.2017.07.016

[3] Marco ML, Heeney D, Binda S, Cifelli CJ, Cotter PD, Foligné B, et al. Health benefits of fermented foods: Microbiota and beyond. Current Opinion in Biotechnology. 2017;**44**:94–102

[4] Huynh N, Van Camp J, Smagghe G, Raes K. Improved release and metabolism of flavonoids by steered fermentation processes: A review. International Journal of Molecular Sciences. 2014;**15**:19369–19388. DOI: 10 .3390/ijms151119369

[5] Srinivasan K. Anti-Inflammatory influences of culinary spices and their bioactives. Food Reviews International. 2022;**38**:1–17. DOI: 10.1080/ 87559129.2020.1839761

[6] Aggarwal BB, Harikumar KB. Potential therapeutic effects of curcumin, the anti-inflammatory agent, against neurodegenerative, cardiovascular, pulmonary, metabolic, autoimmune and neoplastic diseases. The International Journal of Biochemistry & Cell Biology. 2009;**41**:40–59. DOI: 10.1016/ j.biocel.2008.06.010

[7] Mao -Q-Q, Xu X-Y, Cao S-Y, Gan R-Y, Corke H, Beta T, et al. Bioactive compounds and bioactivities of ginger (Zingiber officinale Roscoe). Foods. 2019;**8**:185

[8] Kumar APN, Kumar M, Jose A, Tomer V, Oz E, et al. Major phytochemicals: Recent advances in health benefits and extraction method. Molecules. 2023;**28**:887

[9] D'Archivio M, Filesi C, Benedetto R, Gargiulo R, Giovannini C, Masella R. Polyphenols, dietary sources and bioavailability. Annali Dell'Istituto Superiore Di Sanità. 2007;**43**:348–361

[10] Ajanaku CO, Ademosun OT, Atohengbe PO, Ajayi SO, Obafemi YD, Owolabi OA, et al. Functional bioactive compounds in ginger, turmeric, and garlic. Frontiers in Nutrition. 2022:9. DOI: 10.3389/fnut.2022.1012023

[11] Sawant S, Park H-Y, Sim E-Y, Kim H-S, Choi H-S. Microbial fermentation in food: impact on functional properties and nutritional enhancement-A review of recent developments. Fermentation. 2025;**11**:15. DOI: 10.3390/ fermentation11010015

[12] Gómez-Gaete C, Avendaño-Godoy J, Escobar-Avello D, Campos-Requena VH, Rogel-Castillo C, Estevinho LM, et al. Revolutionizing fruit juice: Exploring encapsulation techniques for bioactive compounds and their impact on nutrition, flavour and shelf life. Food Production, Processing and Nutrition. 2024;**6**:8

[13] Nieto G, Martínez-Zamora L, Peñalver R, Marín-Iniesta F, Taboada-Rodríguez A, López-Gómez A, et al. Applications of plant bioactive compounds as replacers of synthetic

additives in the food industry. Foods. 2024;**13**:47

[14] Ghenabzia I, Hemmami H, Ben Amor I, Zeghoud S, Ben Seghi RB, Hammoudi R. Different methods of extraction of bioactive compounds and their effect on biological activity: A review. International Journal of Secondary Metabolite. 2023;**10**:469–494. DOI: 10.21448/ijsm.1225936

[15] Siddique MAB, Tzima K, Rai DK, Brunton N. Conventional Extraction Techniques for Bioactive Compounds from Herbs and Spices. In: Herbs, Spices and Medicinal Plants. Hossain MB, Brunton NP, Rai DK, editors. 1st. Hoboken, NJ, USA: Wiley; 2020. pp. 69–93. DOI: 10.1002/9781119036685.ch4

[16] Agrawal RS, Nirmal NP. Technologies for Extraction of Bioactive Compounds and Its Applications. In: Sarkar T, Pati S, editors. Bioactive Extraction and Application in Food and Nutraceutical Industries. New York, NY: Springer US; 2024. pp. 1–20. DOI: 10.1007/978-1-0716-3601-5_1

[17] Świder O, Roszko MŁ, Wójcicki M. The inhibitory effects of plant additives on biogenic amine formation in fermented foods – A review. Critical Reviews in Food Science and Nutrition. 2024;**64**:12935–12960. DOI: 10.1080/10408398.2023.2258964

[18] Usman I, Hussain M, Imran A, Afzaal M, Saeed F, Javed M, et al. Traditional and innovative approaches for the extraction of bioactive compounds. International Journal of Food Properties. 2022;**25**:1215–1233

[19] Sánchez-Camargo AP, Montero L, Mendiola JA, Herrero M, and Ibáñez E. Novel Extraction Techniques for Bioactive Compounds from Herbs and Spices. In: Herbs, Spices and Medicinal Plants. Hossain MB, Brunton NP, Rai DK, editors. 1st. Hoboken, NJ, USA: Wiley; 2020. pp. 95–128. DOI: 10.1002/9781119036685.ch5

[20] Mungwari CP, King'ondu CK, Sigauke P, Obadele BA. Conventional and modern techniques for bioactive compounds recovery from plants: Review. Scientific African. 2025;27:e02509. DOI: 10.1016/j.sciaf.2024.e02509

[21] Kim H, Oh J, Hong J. Photodegradation of turmeric oleoresin under fluorescent light and White LED: Impacts on the chemical stability, bioactivity, and photosensitizing property of curcuminoids. Molecules. 2025;**30**:3187. DOI: 10.3390/molecules30153187

[22] Kokkini S, Karousou R, Lanaras T. Essential oils of spearmint (Carvone-rich) plants from the island of Crete (Greece). Biochemical Systematics and Ecology. 1995;**23**:425–430. DOI: 10.1016/0305-1978(95)00021-L

[23] Díaz-Maroto MC, Pérez-Coello MS, González Viñas MA, Cabezudo MD. Influence of drying on the flavor quality of spearmint (Mentha spicata L.). Journal of Agricultural and Food Chemistry. 2003;**51**:1265–1269. DOI: 10.1021/jf020805l

[24] Ghasemzadeh A, Jaafar HZE, Rahmat A. Antioxidant activities, total phenolics and flavonoids content in two varieties of Malaysia young ginger (Zingiber officinale Roscoe). Molecules. 2010;**15**:4324–4333. DOI: 10.3390/molecules15064324

[25] Dai J, Mumper RJ. Plant phenolics: Extraction, analysis and their

antioxidant and anticancer properties. Molecules. 2010;**15**:7313–7352. DOI: 10.3390/molecules15107313

[26] Grigoriadou D, Androulaki A, Psomiadou E, Tsimidou M. Solid phase extraction in the analysis of squalene and tocopherols in olive oil. Food Chemistry. 2007;**105**:675–680. DOI: 10.1016/j.foodchem.2006.12.065

[27] Sultana B, Anwar F, Ashraf M. Effect of extraction solvent/technique on the Antioxidant activity of selected medicinal plant extracts. Molecules. 2009;**14**:2167–2180. DOI: 10.3390/molecules14062167

[28] Amalraj A, Pius A, Gopi S, Gopi S. Biological activities of curcuminoids, other biomolecules from turmeric and their derivatives - A review. Journal of Traditional and Complementary Medicine. 2017;**7**:205–233. DOI: 10.1016/j.jtcme.2016.05.005

[29] Shahbazi R, Sharifzad F, Bagheri R, Alsadi N, Yasavoli-Sharahi H, Matar C. Anti-Inflammatory and immunomodulatory properties of fermented plant foods. Nutrients. 2021;**13**:1516. DOI: 10.3390/nu13051516

[30] Liu Z, Li J-X, Wang C-Z, Zhang D-L, Wen X, Ruan -C-C, et al. Microbial Conversion of Protopanaxadiol-type Ginsenosides by the Edible and Medicinal Mushroom Schizophyllum commune: A green biotransformation strategy. ACS Omega. 2019;**4**:13114–13123

[31] Li Z, Ahn HJ, Kim NY, Lee YN, Ji GE. Korean Ginseng berry fermented by Mycotoxin non-producing Aspergillus niger and Aspergillus oryzae: Ginsenoside analyses and anti-proliferative activities. Biological and Pharmaceutical Bulletin. 2016;**39**:1461–1467. DOI: 10.1248/bpb.b16-00239

[32] Mulyani H, Artanti N, Yuniati R, Yasman Y. Scaling up red Ginger Kombucha Fermentation: Insights into its chemical profile and health-promoting properties. Fermentation. 2025;**11**:128. DOI: 10.3390/fermentation11030128

[33] Yong CC, Yoon Y, Yoo HS, Oh S. Effect of Lactobacillus Fermentation on the anti-inflammatory potential of turmeric. Journal of Microbiology & Biotechnology. 2019;**29**:1561–1569. DOI: 10.4014/jmb.1906.06032

[34] Thierry NN, Léopold TN, Didier M, Moses FMC. Effect of pure culture fermentation on biochemical composition of Moringa oleifera lam leaves powders. Food and Nutrition Sciences. 2013;**4**:851–859. DOI: 10.4236/fns.2013.48111

[35] Jiang W, Qian L, Zhao Y, Lin Y, Yang Y, Shen H, et al. The application of Moringa oleifera Leaf Meal and Its fermentation products in the diet of Megalobrama amblycephala Juveniles. Fermentation. 2023;**9**:577

[36] Adelekan AO, Alamu EA, Daramola BE. Effect of enrichment with turmeric and ginger on some quality characteristics of fermented maize Ogi. Croatian Journal of Food Science and Technology. 2021;**13**:210–220. DOI: 10.17508/CJFST.2021.13.2.11

[37] Schons P, Ries E, Battestin V, Macedo G. Effect of enzymatic treatment on tannins and phytate in sorghum (Sorghum bicolor) and its nutritional study in rats. International Journal of Food Science and Technology. 2011;**46**:1253–1258. DOI: 10.1111/j.1365-2621.2011.02620.x

[38] Mehak M, Deepti S, Gajender Kumar A, Jagdip Singh S, Shilpa V, Deepansh S. Role of lacto-fermentation in reduction of antinutrients in plant-based foods. Journal of Applied Biology & Biotechnology. 2021. doi:10.7324/JABB.2021.9302

[39] Zhang L, Lu J. Rosemary (Rosmarinus officinalis L.) polyphenols and inflammatory bowel diseases: Major phytochemicals, functional properties, and health effects. Fitoterapia. 2024;**177**:106074. DOI: 10.1016/j.fitote.2024.106074

[40] Yadav UCS, Baquer NZ. Pharmacological effects of Trigonella foenum-graecum L. in health and disease. Pharmaceutical Biology. 2014;**52**:243–254. DOI: 10.3109/13880209.2013.826247

[41] Jin H, Seo J-H, Uhm Y-K, Jung C-Y, Lee S-K, Yim S-V. Effect of enrichment with turmeric and ginger on some quality characteristics of fermented maize Ogi. Croatian Journal of Food Science and Technology. 2012;**139**:664–667. DOI: 10.1016/j.jep.2011.11.052

[42] Hussain A, Bose S, Wang J-H, Yadav M, Mahajan G, Kim H. Fermentation, a feasible strategy for enhancing bioactivity of herbal medicines. Food Research International. 2015;81. doi:10.1016/j.foodres.2015.12.026

[43] Cui R, Zhang C, Pan Z-H, Hu T-G, Wu H. Probiotic-fermented edible herbs as functional foods: A review of current status, challenges, and strategies. Comprehensive Reviews in Food Science and Food Safety. 2024;**23**:e13305. DOI: 10.1111/1541-4337.13305

[44] Tai X, Li J, Song J, Zhong B, Li F. Fermented ginseng powder enriched with rare ginsenosides ameliorates high-fat diet-induced obesity by modulating adipogenesis and inflammation. Food and Nutrition Research. 2025;**69**:10.29219/fnr.v69.12230. DOI: 10.29219/fnr.v69.12230

[45] Pérez-Rivero C, López-Gómez JP. Unlocking the Potential of Fermentation in Cosmetics: A Review. Fermentation. 2023;**9**:463. DOI: 10.3390/fermentation9050463

[46] Sahlabgi A, Lupuliasa D, Stanciu G, Lupşor S, Vlaia LL, Rotariu R, et al. The development and comparative evaluation of rosemary hydroalcoholic macerate-based dermatocosmetic preparations: A study on antioxidant, antimicrobial, and anti-inflammatory properties. Gels. 2025;**11**:149

[47] Choi HY, Lee YJ, Kim CM, Lee Y-M. Revolutionizing cosmetic ingredients: Harnessing the power of antioxidants, probiotics, plant extracts, and peptides in personal and skin care products. Cosmetics. 2024;**11**:157. DOI: 10.3390/cosmetics11050157

[48] Ahmad T, Esposito F, Cirillo T. Valorization of Agro-Food by-products: Advancing sustainability and sustainable development goals 2030 through functional compound recovery. Food Bioscience. 2024;**62**:105194. DOI: 10.1016/j.fbio.2024.105194

[49] Martins S, Mussatto SI, Martínez-Avila G, Montañez-Saenz J, Aguilar CN, Teixeira JA. Bioactive phenolic compounds: Production and extraction by solid-state fermentation. A review. Biotechnology Advances. 2011;**29**:365–373. DOI: 10.1016/j.biotechadv.2011.01.008

[50] Sharma R, Garg P, Kumar P, Bhatia SK, Kulshrestha S. Microbial Fermentation and its role in quality improvement of fermented foods. Fermentation. 2020;**6**:106. DOI: 10.3390/fermentation6040106

[51] Awad AM, Kumar P, Ismail-Fitry MR, Jusoh S, Ab Aziz MF, Sazili AQ. Green extraction of bioactive compounds from plant biomass and their application in meat as natural antioxidant. Antioxidants. 2021;**10**:1465. DOI: 10.3390/antiox10091465

[52] Mukherjee A, Gómez-Sala B, O'Connor EM, Kenny JG, Cotter PD. Global Regulatory Frameworks for Fermented Foods: A Review. Frontiers in Nutrition. 2022;9. doi:10.3389/fnut.2022.902642

Chapter 3

Effect of Quercetin on Wound Healing and Tissue Adhesion

Percin Karakol

Abstract

Quercetin, a naturally occurring polyphenolic flavonoid, is widely recognized for its potent antioxidant activity and its role as a secondary plant metabolite. Extensive research has documented its broad spectrum of biological effects, including antiviral, antifungal, antibacterial, antiallergic, cardioprotective, and skin-sensitizing properties. Beyond these established functions, emerging evidence highlights its involvement in wound healing, antiaging processes, and neuroprotection. Due to its hydrophobic nature, quercetin demonstrates slower and limited release compared to water-soluble agents. To overcome these limitations, diverse chemical formulations and advanced delivery technologies have been developed to improve its bioavailability, release kinetics, and stability. Wound healing occurs through three distinct phases, with quercetin exerting particularly pronounced effects during the proliferative and remodeling stages. Prolonged inflammation often results in chronic wounds, bacterial colonization, and tissue degradation; quercetin mitigates these risks by enhancing angiogenesis and epithelialization, thereby accelerating wound closure and reducing infection incidence. Such improvements are critical for lowering treatment costs and facilitating patients' timely return to daily activities. This review synthesizes current knowledge on the molecular and biochemical mechanisms by which quercetin modulates wound healing, emphasizing its therapeutic potential. It further examines strategies to optimize dosing, including nanotechnological approaches that enable the transition from natural compounds to bioengineered therapeutic platforms. The integration of cost-effective technologies with natural resources may not only provide individual psychosocial benefits but also contribute to broader public health and socioeconomic advancements.

Keywords: quercetin, wound healing, oxidative stress, natural sources, flavonoids

1. Introduction

A wound is broadly defined as a disruption of skin integrity, ranging from superficial epidermal injury to deeper involvement of bone, muscle, tendon, and neurovascular structures. The most common etiologies include trauma, burns, neoplasms, circulatory disorders, infections, diabetes, and immune dysfunction. Wound healing represents a dynamic and highly regulated biological process that unfolds through three overlapping phases: hemostasis/inflammation, proliferation, and remodeling. Despite its complexity, normal healing proceeds in a coordinated manner, resembling a biological symphony in which mediators, cytokines, complement factors, and cellular players such as endothelial cells, keratinocytes, and fibroblasts interact to restore tissue integrity [1].

Failure of wounds to heal or progression to chronic states imposes substantial morbidity on affected individuals and generates significant socioeconomic and healthcare burdens. Consequently, elucidating the molecular and cellular mechanisms underlying wound repair and identifying bioactive compounds derived from natural sources has become a priority in contemporary biomedical research.

Current wound management strategies encompass topical and systemic agents, specialized dressings, epidermal and dermal substitutes, and, in refractory cases, surgical intervention. Traditional medicine has long employed plant-based therapies for wound healing, with phytochemicals, particularly flavonoids, receiving increasing attention due to their antioxidant, antiinflammatory, and anticoagulant properties. Flavonoids have demonstrated broad therapeutic potential, including antioxidative, neuroprotective, cardioprotective, anticancer, antiviral, antibacterial, and antiallergic effects [2].

Among these, quercetin, a polyhydroxyflavone abundantly present in nature and the human diet, exhibits potent biological activity primarily through the modulation of reactive oxygen species (ROS) signaling. Experimental and emerging clinical evidence support quercetin's role in wound healing, underscoring its promise as a nature-inspired therapeutic agent [3-5].

1.1 Wound healing: Phases and clinical implications

Wound formation follows a complex yet sequential biological pathway and may arise either as a primary event or as a secondary manifestation of underlying pathology. Irrespective of the etiology, the initial physiological response is characterized by vasoconstriction and coagulation to minimize blood loss. This is rapidly followed by complement activation and the initiation of an inflammatory cascade, which recruits neutrophils and macrophages to the wound site [6]. Prolonged or dysregulated inflammation, particularly when compounded by microbial contamination, can impair healing and, in severe cases, lead to life-threatening complications [1].

At this stage, therapeutic interventions aimed at modulating inflammation have gained increasing attention, with particular focus on agents derived from botanical and synthetic sources. The subsequent phases of wound healing – parenchymal cell proliferation, collagen deposition, and extracellular matrix remodeling – are frequently delayed in chronic wounds. A major contributor to this delay is the excessive accumulation of ROS, which disrupts the transition from the inflammatory to the proliferative phase [7]. While physiological levels of ROS are essential for antimicrobial defense and intracellular signaling, sustained elevation induces oxidative stress, impairs angiogenesis, and perpetuates tissue damage [8].

A comprehensive understanding of these molecular disruptions is critical for the development of targeted therapeutic strategies. While physiological ROS activation supports redox balance and homeostasis, prolonged exposure may damage cellular components and contribute to chronic inflammation and disease. Among promising candidates, flavonoid based agents such as quercetin have attracted considerable interest due to their potent antioxidant and antiinflammatory properties, which enable modulation of ROS-mediated pathways and support the restoration of normal wound-healing dynamics.

1.2 Flavonoids: Biological activities and mechanisms of action

Flavonoids represent a diverse group of polyphenolic compounds abundantly distributed in fruits, vegetables, and medicinal plants. Structurally defined by a C6–C3–C6 backbone, they are subclassified into flavonols, flavones, flavanones, isoflavones, and anthocyanidins [5, 9]. Among these, flavonols such as quercetin, kaempferol, myricetin, and fisetin have attracted considerable attention due to their broad pharmacological activities. The biological effects of flavonoids can be categorized into several major mechanisms:

1. **Antioxidant activity**
   - Scavenging of ROS
   - Upregulation of endogenous antioxidant enzymes (e.g., superoxide dismutase [SOD], catalase)
   - Inhibition of lipid peroxidation
   - Protection of cellular components against oxidative damage [4, 10]
2. **Antiinflammatory effects**
   - Downregulation of NF-κB and COX-2 signaling pathways
   - Suppression of proinflammatory cytokines (TNF-α, IL-1β, IL-6)
   - Reduction of leukocyte infiltration and edema formation [11]
3. **Antimicrobial and antiviral properties**
   - Disruption of microbial cell membranes
   - Inhibition of viral replication enzymes
   - Enhancement of host immune responses
   - Potential application in infectious disease management [12]
4. **Antiapoptotic and cytoprotective roles**
   - Regulation of apoptotic signaling via Bcl-2/Bax modulation
   - Inhibition of caspase activation
   - Promotion of cell survival under stress conditions [13]

5. **Angiogenesis and tissue regeneration**

   - Stimulation of vascular endothelial growth factor (VEGF) expression
   - Induction of endothelial cell proliferation
   - Support for neovascularization and tissue repair is particularly relevant in wound healing [14]

1.3 Quercetin: Structure, biological activities, and role in wound healing

1. **Structural characteristics**

   Quercetin is a polyhydroxyflavone composed of two benzene rings linked by a three-carbon bridge, forming a closed pyran ring. Its molecular structure contains five hydroxyl groups, each of which can undergo glycosylation, giving rise to multiple quercetin glycoside derivatives with distinct pharmacokinetic properties [15]. Quercetin is abundantly present in apples, citrus fruits, onions, grapes, plums, tomatoes, buckwheat, legumes, tea, coriander, *Centella asiatica*, *Krishna jiraka*, *Ginkgo biloba*, and garden cress [16].

2. **Biological activities relevant to wound healing**

   Quercetin exerts a wide range of biological effects that directly contribute to tissue repair and regeneration:

   - **Antiallergic activity:** stabilization of mast cell membranes, preventing degranulation and histamine release.
   - **Antioxidant properties:** scavenging of free radicals such as hydroxyl, alkoxyl, and lipid peroxyl species, thereby reducing oxidative stress [17].
   - **Enzyme inhibition and chelation:** inhibition of lipoxygenase and cyclooxygenase enzymes, along with iron-chelating activity, contributes to antiinflammatory and antioxidant effects [18].
   - **Immunomodulation:** regulation of cytokine production and macrophage activity, supporting immune balance.
   - **Pharmacokinetics:** despite potent bioactivity, quercetin demonstrates low oral bioavailability due to poor solubility and extensive first-pass hepatic metabolism [19].

   During tissue injury, endogenous antioxidant enzymes such as catalase, SOD, and glutathione peroxidase are activated to neutralize excessive ROS generated at the wound site [20–22]. Quercetin enhances this defense by limiting ROS accumulation and modulating macrophage activity, thereby facilitating the resolution of inflammation and promoting the transition to the proliferative phase of wound healing [1] (**Table 1**).

Molecular mechanisms	Biological effects
• NF–kB inhibition	• Antiinflammatory
• SOD and catalase boosting	• Protects against oxidative stress
• MAPK and PI3K/Akt modulation	• Proliferation
• TGF-B1 activation	• Fibroblast proliferation
• Collagen gene expression	• ECM remodeling
• VEGF increase	• Angiogenesis

Table 1.
Molecular structure and biological functions of quercetin in wound healing.

3. **Systemic pharmacological activities**
   Beyond wound healing, quercetin exhibits a broad spectrum of pharmacological effects that reinforce its therapeutic versatility:

- **Cardioprotective effects:** improvement of endothelial function, reduction of blood pressure, and inhibition of LDL oxidation, thereby lowering the risk of atherosclerosis and ischemic injury [23].

- **Neuroprotective actions:** the ability to cross the blood–brain barrier and modulate neuroinflammatory pathways, offering protection against neurodegenerative disorders such as Alzheimer's and Parkinson's disease [24].

- **Antidiabetic properties:** enhancement of insulin sensitivity, regulation of glucose metabolism, and protection of pancreatic β-cells from oxidative damage [25].

- **Anticancer potential:** induction of apoptosis, inhibition of cell proliferation, and interference with angiogenesis in various cancer models, including breast, prostate, and colon cancers [26].

- **Antiviral and antiparasitic activity:** inhibition of viral replication enzymes and suppression of parasitic growth, supporting its potential as adjunctive therapy in infectious diseases [27].

4. **Challenges and advances in delivery systems**
   Despite its pleiotropic actions, quercetin's hydrophobic nature limits aqueous solubility, resulting in poor bioavailability and stability. To overcome these challenges, diverse chemical formulations and nanotechnology-based delivery systems have been developed to enhance its pharmacokinetic profile, stability, and targeted release. These systemic effects are largely mediated through the modulation of oxidative stress, inflammatory signaling, and cellular homeostasis.

5. **Clinical relevance**
   The pleiotropic biological activities of quercetin underscore its relevance not only in localized tissue repair but also in broader clinical applications. Its integration into nanotechnological platforms may facilitate translation from

natural compound to bioengineered therapeutic agent, thereby expanding its utility in wound healing and systemic disease management.

2. Results

2.1 Quercetin's therapeutic impact on wound healing

1. **Hemostasis and inflammation phase**
   Following tissue injury, wound healing begins with vasoconstriction and platelet aggregation, leading to clot formation and hemostasis. This initial response is rapidly followed by the recruitment of immune cells, primarily neutrophils and macrophages, mediated through the release of inflammatory cytokines and chemokines such as TNF-α, IL-1β, and IL-6. Quercetin modulates this phase by downregulating NF-κB signaling and suppressing proinflammatory mediators. Through this mechanism, it prevents prolonged inflammation and promotes a controlled immune response, thereby reducing the risk of chronic wound development [1, 28]. TNF-α, IL-1β, and NF-κB are critical biomarkers of wound healing and play a central role in regulating the inflammatory response. TNF-α and IL-1β promote tissue clearance and defense by triggering neutrophil and macrophage migration in the early phase, while NF-κB controls the inflammatory cascade as a transcriptional activator of these cytokines. At physiological levels, these processes facilitate the transition to the proliferative phase; however, excessive or prolonged activation results in chronic inflammation, increased oxidative stress, fibroblast dysfunction, and delayed collagen synthesis. Clinically, the levels of these markers are important as measurable parameters for predicting wound prognosis and assessing response to treatment. In addition, antiinflammatory and antioxidant agents, especially flavonoids such as quercetin, are of great clinical value in translational wound care, as they suppress these pathways, thereby controlling inflammation and accelerating healing (**Table 2**).

2. **Proliferation phase**
   The proliferative phase is characterized by fibroblast proliferation, keratinocyte migration, angiogenesis, and granulation tissue formation. While ROS

Inflamation	• Modulates inflammation • Reduces oxidative stress • Inhibits-NF–kB pathway
Proliferation	• Promotes fibroblast proliferation • Enhances granulation tissue formation • Reduces ROS accumulation
Remodeling	• Stimulates collagen synthesis • Improves ECM organization • Enhances dermal remodeling

Table 2.
Quercetin's molecular actions across wound-healing phases.

are essential for cellular signaling, excessive accumulation can impair cellular function and delay repair. Acting as a potent antioxidant, quercetin scavenges ROS and enhances the activity of endogenous enzymes such as catalase and SOD. This dual action reduces oxidative stress, supports cellular proliferation, and facilitates extracellular matrix deposition, thereby accelerating tissue regeneration [5].

3. **Remodeling phase**
   The final stage of wound healing involves ECM reorganization, collagen maturation, and wound contraction. Fibroblasts play a central role in synthesizing and aligning collagen fibers to restore tensile strength and structural integrity. Quercetin promotes collagen synthesis and improves ECM remodeling by modulating fibroblast activity and reducing the expression of matrix-degrading enzymes. These effects contribute to enhanced dermal regeneration, improved scar quality, and restoration of functional tissue architecture [15, 16].

3. Discussion

Collectively, the integration of quercetin into advanced biomaterial platforms, such as hydrogels, nanocrystals, and lipid-based vesicles, represents a significant leap toward translational wound care. These systems not only overcome quercetin's inherent limitations in solubility and stability but also enable targeted delivery to wound sites, maximizing therapeutic outcomes. Future clinical trials are essential to validate these findings and establish standardized protocols for quercetin-based interventions across diverse wound etiologies.

Quercetin, a polyphenolic flavonoid abundantly found in various medicinal plants, has emerged as a promising candidate in wound healing due to its potent antioxidant, antiinflammatory, and antimicrobial properties. These bioactivities are particularly relevant in chronic wounds, where prolonged oxidative stress and dysregulated inflammation impair tissue regeneration and delay healing.

Recent studies have emphasized quercetin's ability to modulate the wound microenvironment at multiple levels. Nazarnezhad et al. [29] demonstrated that quercetin-loaded biocomposites significantly enhance skin defense mechanisms by neutralizing oxidizing agents and inflammatory triggers. Their review highlights quercetin's role in stabilizing cellular membranes, reducing lipid peroxidation, and promoting fibroblast migration – key events in the proliferative phase of healing.

In a complementary study, Nayak et al. [30] developed a quercetin nanocrystal-loaded alginate hydrogel patch that improved solubility and sustained release, resulting in accelerated wound closure and enhanced histological outcomes in preclinical models. The hydrogel's biocompatibility and ability to maintain a moist wound environment further contributed to its therapeutic efficacy.

Moreover, glycerohyalurosome-based delivery systems have shown promise in optimizing quercetin's bioavailability. Aghajanloo et al. [31] formulated a quercetin-loaded hydrogel using hyaluronic acid-modified glycerosomes, which facilitated deeper tissue penetration and prolonged antioxidant activity. This approach not only improved epithelialization but also reduced inflammatory cell infiltration, suggesting a dual-phase benefit in both the early and late stages of healing.

Flavonoids, and particularly quercetin, have attracted considerable attention for their integration into nanotechnology-based delivery systems in wound healing, offering both distinct advantages and notable limitations. On the one hand, nanocarriers such as liposomes, polymeric nanoparticles, nanoemulsions, and hydrogels significantly enhance quercetin's aqueous solubility, bioavailability, and pharmacokinetic stability, enabling controlled release and targeted delivery to wound sites. These systems also prolong antioxidant activity, sustain antiinflammatory effects, and create biocompatible environments that support fibroblast proliferation, collagen deposition, and angiogenesis, thereby accelerating tissue regeneration. On the other hand, the complexity of formulation, potential long-term biocompatibility concerns, high production costs, and limited clinical validation remain important challenges that hinder widespread adoption. Thus, while nanotechnological platforms provide a promising translational pathway for maximizing quercetin's therapeutic potential, future research must address safety, scalability, and regulatory issues to ensure its effective clinical application in wound management.

These findings align with earlier reports by Karakol et al. [32, 33], who emphasized quercetin's role in regulating mast cell activity, enhancing collagen synthesis, and mitigating ischemia-reperfusion injury through the upregulation of endogenous antioxidant enzymes such as catalase and SOD.

Chronic wounds represent a significant global health burden, often requiring a multidisciplinary approach due to their complex pathophysiology and resistance to conventional therapies. In this context, every novel therapeutic strategy carries substantial clinical relevance. Quercetin, a naturally occurring polyhydroxyflavonoid, has garnered attention not only for its oral and topical applications but also for its integration into regenerative medicine and tissue engineering platforms.

The wound-healing potential of quercetin is primarily attributed to its potent antioxidant capacity and its ability to modulate inflammatory responses. Experimental studies have demonstrated that quercetin suppresses the expression of proinflammatory cytokines, promotes fibroblast proliferation, and enhances collagen synthesis – mechanisms that are particularly beneficial in the management of chronic wounds [34–36].

Despite its promising bioactivity, quercetin's poor aqueous solubility and limited bioavailability remain major obstacles to its clinical translation. To overcome these limitations, various nanoformulations, including liposomes, nanoemulsions, and polymeric nanoparticles, have been developed to facilitate controlled release and targeted tissue delivery. These advanced delivery systems significantly enhance the therapeutic efficacy and safety profile of quercetin in wound care applications [37].

The use of natural biomaterials in combination with quercetin also offers advantages in terms of environmental sustainability and cost-effectiveness. Integrating quercetin into nanotechnology-based platforms not only addresses individual health challenges but also contributes to reducing the burden on healthcare systems. Formulation strategies that enable quercetin to reach therapeutic concentrations require interdisciplinary collaboration and offer fertile ground for future research.

Recent studies have explored the incorporation of quercetin into biocompatible and bioinert matrices such as hydrogels, nanojels, and nanoparticles. For instance, antioxidant hydrogels composed of poly(aspartic acid) and carboxymethylcellulose loaded with quercetin have shown efficacy as burn wound dressings [38]. Similarly, nanocrystal formulations have demonstrated enhanced topical activity and improved skin penetration.

Given its hydrophobic nature, quercetin necessitates the use of specialized carriers for pharmaceutical applications. Biomaterials have proven effective not only in wound healing but also in postburn tissue regeneration. Collagen-based matrices, in particular, have been shown to improve quercetin release kinetics while simultaneously leveraging the regenerative properties of extracellular matrix components [39, 40].

ECM plays a pivotal role throughout all phases of wound healing, serving as a dynamic reservoir for signaling molecules and cellular interactions. Among the key cellular players, fibroblasts are central to ECM remodeling and tissue regeneration. Quercetin has been shown to accelerate wound healing in diabetic models by mitigating oxidative stress and stimulating fibroblast-driven proliferation [34, 35].

Furthermore, quercetin exerts antiinflammatory effects by modulating macrophage polarization, shifting the phenotype from proinflammatory M1 to reparative M2, thereby enhancing tissue repair [35]. Beyond inflammation, quercetin influences the remodeling phase by attenuating fibrotic signaling pathways and promoting angiogenic factors such as VEGF and TGF-β1, which are essential for scar modulation and tissue integrity [36]. These biochemical changes translate into enhanced fibroblast proliferation, improved collagen deposition, and accelerated epithelialization, which can be assessed through histological scoring, tensile strength measurements, and wound closure rates in experimental models. Collectively, these quantifiable outcomes underscore the clinical relevance of quercetin as a multifunctional agent in wound repair, providing a robust evidence base for its integration into translational wound care strategies.

In addition, combinatorial approaches involving quercetin and other phytochemicals, such as curcumin, have demonstrated synergistic effects in antibacterial activity, radical scavenging, and dermal fibroblast migration, further supporting its role in integrative wound therapy [41].

Collectively, the integration of quercetin into advanced biomaterial platforms, such as hydrogels, nanocrystals, and lipid-based vesicles, represents a significant leap toward translational wound care. These systems not only overcome quercetin's inherent limitations in solubility and stability but also enable targeted delivery to wound sites, maximizing therapeutic outcomes. Future clinical trials are essential to validate these findings and establish standardized protocols for quercetin-based interventions across diverse wound etiologies.

4. Conclusion

This review has critically examined the chemical and molecular mechanisms through which quercetin contributes to wound healing, highlighting its multifaceted biological activities. As a potent antioxidant within the flavonoid family, quercetin exerts pronounced effects during the proliferative and remodeling phases of tissue repair. Its capacity to modulate inflammatory responses, enhance fibroblast proliferation, and stimulate collagen synthesis underscores its therapeutic potential in regenerative medicine and wound management.

Despite these promising properties, quercetin's poor aqueous solubility and limited bioavailability remain significant barriers to clinical translation. To overcome these limitations, nanotechnology-based delivery systems and advanced formulation strategies have been developed to improve controlled release, stability, and targeted delivery. The transformation of natural compounds into bioactive materials through such technologies offers scalable, cost-effective solutions with both individual and societal benefits.

Compared to flavonoids such as catechins, rutin, or hesperidin, which act predominantly in single domains (e.g., antioxidation or vascular modulation), quercetin exerts multiphase effects spanning inflammation, proliferation, and remodeling. Furthermore, quercetin has been successfully integrated into advanced biomaterial platforms, including hydrogels, nanocrystals, and lipid-based vesicles, overcoming its inherent solubility and stability limitations. This translational readiness sets quercetin apart from other flavonoids, which have not yet achieved comparable integration into wound-specific delivery systems.

The future prospects of natural compounds, particularly flavonoids such as quercetin, are increasingly recognized in the development of novel pharmaceuticals and functional nutraceuticals. Given its pleiotropic biological activities, quercetin represents a valuable candidate for therapeutic innovation in wound care. However, widespread clinical adoption requires comprehensive studies to elucidate its pharmacodynamics, safety profile, drug interactions, and optimal dosing regimens. The advancement of high-bioavailability formulations and biocompatible scaffolds remains essential to fully realize its clinical potential.

In the broader context of modern biomedical challenges, quercetin, alongside other bioactive molecules such as terpenes, polyphenols, and triterpenoids, should be considered among the leading natural agents for accelerating wound healing. Furthermore, semisynthetic derivatives and hybrid biomaterials warrant exploration to enhance therapeutic efficacy and material performance.

Ultimately, future research should prioritize validating quercetin's clinical efficacy across diverse formulations and wound types, supported by multidisciplinary approaches integrating phytochemistry, materials science, and nanomedicine. Such efforts may pave the way for a paradigm shift in chronic wound management, positioning quercetin as a key molecule in both basic and translational biomedical research.

Author details

Percin Karakol
Department of Plastic, Reconstructive, and Aesthetic Surgery, Basakşehir Cam and Sakura City Hospital, University of Health Sciences, Istanbul, Türkiye

*Address all correspondence to: ppercin@gmail.com

References

[1] Eming SA, Martin P, Tomic-Canic M. Wound repair and regeneration: Mechanisms, signaling, and translation. Science Translational Medicine. 2014; **6**(265):265sr6. DOI: 10.1126/scitranslmed.3009337

[2] Jomova K, Raptova R, Alomar SY, Alwasel SH, Nepovimova E, Kuca K, Valko M. Reactive oxygen species, toxicity, oxidative stress, and antioxidants: Chronic diseases and aging. Archives of Toxicology. 2023;**97**(10):2499–2574. DOI: 10.1007/s00204-023-03562-9

[3] McKay TB, Emmitte KA, German C, Karamichos D. Quercetin and related analogs as therapeutics to promote tissue repair. Bioengineering (Basel). 2023;**10**(10):1127. DOI: 10.3390/bioengineering10101127

[4] Miean KH, Mohamed S. Flavonoid (Myricetin, quercetin, kaempferol, luteolin, and apigenin) content of edible tropical plants. Journal of Agricultural and Food Chemistry. 2001;**49**:3106–3112

[5] Jomova K, Alomar SY, Valko R, Liska J, Nepovimova E, Kuca K, Valko M. Flavonoids and their role in oxidative stress, inflammation, and human diseases. Chemico-Biological Interactions. 2025;**413**:111489. DOI: 10.1016/j.cbi.2025.111489

[6] Chittasupho C, Junmahasathien T, Chalermmongkol J, Wongjirasakul R, Leesawat P, Okonogi S. Suppression of intracellular reactive oxygen species in human corneal epithelial cells via quercetin nanoparticles and epigallocatechin gallate in situ thermosensitive gel. Pharmaceuticals. 2021;**14**:679

[7] Sen CK. Wound healing essentials: Let there be oxygen. Wound Repair and Regeneration. 2009;**17**(1):1–18. DOI: 10.1111/j.1524-475X.2008.00436.x

[8] Roy S, Khanna S, Nallu K, Hunt TK, Sen CK. Dermal wound healing is subject to redox control. Molecular Therapy. 2006;**13**(1):211–220. DOI: 10.1016/j.ymthe.2005.07.684

[9] Mahmud AR, Ema TI, Siddiquee MF-R, et al Natural flavonols: Actions, mechanisms, and potential therapeutic utility. Beni-Suef University Journal of Basic and Applied Sciences. 2023;**12**:47

[10] Jomova K, Valko M. Advances in metal-induced oxidative stress and human disease. Toxicology. 2011; **283**(2–3):65–87. DOI: 10.1016/j.tox.2011.03.001

[11] Stachelska MA, Karpiński P, Kruszewski B. Biological properties of flavonoids and their role in prevention of metabolic, cancer and neurodegenerative diseases. Applied Sciences. 2025;**15**(19):10840. DOI: 10.3390/app151910840

[12] Cushnie TP, Lamb AJ. Antimicrobial activity of flavonoids. International Journal of Antimicrobial Agents. 2005;**26**(5):343–356. DOI: 10.1016/j.ijantimicag.2005.09.002

[13] Spencer JP. Flavonoids: Modulators of brain function? British Journal of Nutrition. 2008;**99**(Suppl 1):ES60–ES77. DOI: 10.1017/S0007114508006974

[14] Weng CJ, Yen GC. Chemopreventive effects of dietary phytochemicals against cancer invasion

and metastasis: phenolic acids, monophenol, polyphenol, and their derivatives. Cancer Treat Rev. 2012;**38**(1):76–87. DOI: 10.1016/j.ctrv.2011.03.001

[15] Wang G, Wang Y, Yao L, et al Pharmacological activity of quercetin: An updated review. Evidence-Based Complementary and Alternative Medicine. 2022;**2022**: 3997190

[16] Bhagwat S, Haytowitz DB, Holden JM. USDA Database for the Flavonoid Content of Selected Foods, Release 3. Beltsville, MD: US Department of Agriculture; 2011

[17] Polerà N, Badolato M, Perri F, Carullo G, Aiello F. Quercetin and its natural sources in wound healing management. Current Medicinal Chemistry. 2019;**26**(31):5825–5848. DOI: 10.2174/0929867325666180713150626

[18] Morel I, Lescoat G, Cogrel P, et al. Antioxidant and iron-chelating activities of catechin, quercetin, and diosmetin. Biochemical Pharmacology 1993;**45**:13–19

[19] Chen X, Yin OQ, Zuo Z, Chow MS. Pharmacokinetics and modeling of quercetin and metabolites. Pharmaceutical Research. 2005;**22**(6):892–901. DOI: 10.1007/s11095-005-4584-1

[20] Aghababaei F, Hadidi M. Recent advances in potential health benefits of quercetin. Pharmaceuticals (Basel). 2023;**16**(7):1020. DOI: 10.3390/ph16071020

[21] Mirza MA, Mahmood S, Hilles AR, et al. Quercetin as a therapeutic product: Evaluation of pharmacological action and clinical applications. Pharmaceuticals (Basel). 2023;**16**(11):1631. DOI: 10.3390/ph16111631

[22] Islam MS, Quispe C, Hossain R, et al Neuropharmacological effects of quercetin: A literature-based review. Frontiers in Pharmacology. 2021;**12**:665031

[23] Pérez-Vizcaíno F, Duarte J. Flavonols and cardiovascular disease. Molecular Aspects of Medicine. 2010;**31**(6):478–494. DOI: 10.1016/j.mam.2010.09.002

[24] Nabavi SF, Braidy N, Gortzi O, et al Neuroprotective effects of quercetin: From chemistry to medicine. Neurochemistry International. 2015;**89**:140–150

[25] Eid HM, Nachar A, Thong F, Sweeney G, Haddad PS. The molecular basis of the antidiabetic action of quercetin in cultured skeletal muscle cells and hepatocytes. Pharmacogn Mag. 2015;**11**(41):74–81. DOI: 10.4103/0973-1296.149708

[26] Li XR, Qi L, Zhang XW, Wei C, Yu B, Pei TL. Quercetin and Nano-Derivatives: Potential and Challenges in Cancer Therapy. Int J Nanomedicine. 2015;**20**:6701–6720. DOI: 10.2147/IJN.S509877

[27] Colunga Biancatelli RM, Berrill M, Catravas JD, Marik PE. Quercetin and vitamin C: Synergistic therapy for SARS-CoV-2 disease. Frontiers in Immunology. 2020;**11**:1451. DOI: 10.3389/fimmu.2020.01451

[28] Akhtari N, Ahmadi M, Kiani Doust Vaghe Y, et al. Natural agents as wound-healing promoters. Inflammopharmacology. 2024;**32**(1):101–125. DOI: 10.1007/s10787-023-01318-6

[29] Nazarnezhad S, Nasiri SN, Gorgani S, et al. Quercetin-loaded biocomposites for wound healing applications. Phytochemical Reviews. 2025. DOI: 10.1007/s11101-025-10185-8

[30] Nayak M, Kumar V, Banerjee D, et al. Quercetin nanocrystal-loaded alginate hydrogel patch for wound healing. Journal of Materials Chemistry B. 2025;**13**(5):1690–1703. DOI: 10.1039/d4tb01699h

[31] Abualsunun W, Alahmadi AA, Eshmawi BA, et al Optimized quercetin-loaded glycerohyalurosome hydrogel: An innovative nanoplatform for enhanced wound healing. Future Journal of Pharmaceutical Sciences. 2025;**11**:56

[32] Karakol P, Bozkurt M, Sezgic M, Uslu C, Balikci T. The effect of antioxidants on ischemia-reperfusion injury in flap surgery. In: Antioxidants. London: IntechOpen; 2019. DOI: 10.5772/intechopen.85500

[33] Karakol P, Kapi E. Use of Selected Antioxidant-Rich Spices and Herbs in Foods. In: London: IntechOpen; 2021. DOI: 10.5772/intechopen.96136

[34] Kant V, Sharma M, Jangir BL, Kumar V. Acceleration of wound healing by quercetin in diabetic rats requires mitigation of oxidative stress and stimulation of the proliferative phase. Biotechnic & Histochemistry. 2022;**97**(6):461–472

[35] Fu J, Huang J, Lin M, Xie T, You T. Quercetin promotes diabetic wound healing via switching macrophages from M1 to M2 polarization. Journal of Surgical Research. 2019;**246**:213–223

[36] Gopalakrishnan A, Ram M, Kumawat S, Tandan S, Kumar D. Quercetin accelerated cutaneous wound healing in rats by increasing levels of VEGF and TGF-βd1. Indian Journal of Experimental Biology. 2016;**54**:187–195

[37] Shegokar R, Müller RH. Nanocrystals: Industrially feasible multifunctional formulation technology for poorly soluble actives. International Journal of Pharmaceutics. 2010; **399**(1–2):129–139

[38] Zhu J, Zhang K, Zhang Y, Zhou C, Cui Z, Li W, et al. Antioxidant hydrogel from poly(aspartic acid) and carboxymethylcellulose with quercetin loading as burn wound dressing. International Journal of Biological Macromolecules 2024;**282**:137323

[39] Elfawy LA, Ng CY, Amirrah IN, Mazlan Z, Wen APY, Fadilah NIM, et al. Sustainable approach of functional biomaterials–tissue engineering for skin burn treatment: A comprehensive review. Pharmaceuticals (Basel) 2023;**16**:701

[40] Gomathi K, Gopinath D, Rafiuddin Ahmed M, Jayakumar R. Quercetin incorporated collagen matrices for dermal wound healing processes in rat. Biomaterials. 2003;**24**(16):2767–2772

[41] Chittasupho C, Manthaisong A, Okonogi S, Tadtong S, Samee W. Effects of quercetin and curcumin combination on antibacterial, antioxidant, in vitro wound healing and migration of human dermal fibroblast cells. International Journal of Molecular Sciences. 2021; **23**(1):142. DOI: 10.3390/ijms23010142

Chapter 4

High-resolution LC-MS/MS-based Metabolite Profiling of Medicinal Herbs and Spices: Innovations in Untargeted and Targeted Phytochemical Characterization

Yodying Yingchutrakul and Sucheewin Krobthong

Abstract

High-resolution liquid chromatography–tandem mass spectrometry (LC-MS/MS) has reshaped phytochemical characterization by enabling the sensitive and structurally resolved profiling of metabolites in medicinal herbs and spices. This chapter outlines methodological innovations spanning sample preparation, class-aware extraction strategies, and chromatographic modes optimized for complex botanical matrices. Advances in ionization chemistry, adaptive acquisition logic, and ion-mobility integration have increased spectral richness while improving discrimination among isomeric or low-abundance metabolites. Post-processing frameworks that combine spectral libraries, *in silico* fragmentation, retention-time prediction, and CCS filtering now support transparent, multi-evidence annotation consistent with modern confidence-level standards. Case studies illustrate how LC-MS/MS enhances botanical authentication, quality assessment, and mechanistic interpretation relevant to pharmacological and nutraceutical applications. By uniting high-resolution analytics with open spectral resources and machine-assisted annotation, LC-MS/MS provides a scalable and reproducible foundation for research, regulatory standardization, and functional-food innovation.

Keywords: orbitrap, ion-mobility mass spectrometry, spectral annotation, botanical authentication, natural product, data-dependent acquisition, data-independent acquisition, parallel reaction monitoring, multiple reaction monitoring

1. Introduction

High-resolution liquid chromatography–tandem mass spectrometry (LC–MS/MS) has become the cornerstone of phytochemical analysis because accurate mass measurements, isotopic fidelity, and detailed fragmentation collectively enable the confident detection and annotation of thousands of metabolites across structurally

IntechOpen

diverse herbal matrices [1, 2]. These analytical capabilities not only enhance metabolite detection across diverse phytochemical classes but also facilitate comprehensive profiling strategies that clarify chemically meaningful relationships within botanical matrices. Such strength places LC–MS/MS at the center of modern phytochemical science, establishing a framework that links analytical precision with biological interpretation. This emphasis naturally introduces the importance of workflow design, which ultimately governs the quality and interpretability of phytochemical datasets.

The conceptual shift from instrument-centric analysis to workflow-centric logic is central to current metabolomics practice because study design, extraction chemistry, chromatographic selectivity, ionization strategy, acquisition settings, and post-processing pipelines jointly determine analytical outcomes rather than any single component alone [3]. Medicinal herbs and spices contain chemically dense and structurally diverse metabolites – including flavonoids, phenolic acids, alkaloids, terpenoids, and glycosides – requiring methodological consistency from sample preparation through data interpretation. Because of this complexity, workflow standardization has become essential for ensuring that phytochemical fingerprints reflect biological reality rather than technical variability. This standardization also motivates the development of reproducible computational ecosystems and transparent annotation frameworks.

In parallel with these methodological advances, LC–MS/MS has enabled botanicals to be mapped with unprecedented structural detail, supporting applications ranging from taxonomic authentication to mechanistic inference for bioactive compounds [4, 5]. The ability to differentiate isomeric flavonoids, terpenoid subclasses, or conjugated phenolics is particularly critical in herbal science, where minor structural variations can produce major differences in pharmacological activity. Improved chromatographic resolution and optimized ionization chemistries have therefore become essential for ensuring high-quality spectral acquisition across wide chemical landscapes. These analytical considerations reinforce the importance of informatics tools capable of exploiting the structural richness embedded within LC–MS/MS data.

Advances in open-source computational environments now support reproducible peak detection, spectral deconvolution, and annotation across laboratories and acquisition modes [6, 7]. Tools such as MZmine and MS-DIAL streamline data processing for both data-dependent acquisition (DDA) and data-independent acquisition (DIA) workflows, while enabling the export of interoperable spectral datasets essential for phytochemical annotation. Deep-learning models such as MS2DeepScore and MS2Query further extend structural inference into regions of chemical space where reference libraries remain sparse, enhancing the proportion of annotated features in complex herbal extracts [8, 9]. These developments underline the increasingly critical role of computational innovation in natural-product chemistry, where interpretability often depends on integrating spectral, structural, and class-level evidence.

To ensure that phytochemical discoveries remain reproducible and cumulative, multi-evidence annotation frameworks have become essential. These frameworks combine spectral libraries, *in silico* fragmentation models, retention-time prediction, ion-mobility collision cross-section constraints, and class-level annotations to generate explicit confidence levels aligned with MSI and Schymanski criteria [10, 11]. Such transparent reporting is particularly important in phytochemical characterization, where closely related natural products often co-occur and share overlapping fragmentation behaviors. The emergence of pan-repository infrastructures and FAIR-oriented deposition practices further enables botanicals to be compared across studies and instrumentation platforms, strengthening the foundations of evidence-based herbal science.

Global resources, including machine-readable spectra, community-curated databases, and manufacturer-agnostic search languages, have expanded the scale at which LC–MS/MS data can be interpreted [12, 13]. These resources allow phytochemical datasets to be interrogated across laboratories, regions, and species, thereby accelerating the identification of chemotaxonomic markers, bioactive constituents, and adulteration signatures in medicinal herbs and spices. As these repositories continue to evolve, phytochemical profiling is shifting toward an analyzable and cumulative discipline that supports regulatory standardization, functional-food innovation, and the rediscovery of traditional medicinal knowledge. Collectively, these advances provide the foundation for the remainder of this chapter, which evaluates the methodological, technological, and interpretive developments that characterize contemporary LC–MS/MS-based phytochemical analysis.

2. Study design, sampling, and extraction: Building analytical fitness-for-purpose

Robust study design is the most effective determinant of reproducibility in LC–MS/MS profiling of herbal matrices because biological replication, blocking, randomized injection order, and pooled quality-control (QC) samples permit explicit modeling of drift and batch effects over long sequences [14]. To improve readability, the sequence of technical terms was separated into shorter explanatory units, emphasizing that each design element independently contributes to the statistical control of variability. QC-based robust LOESS signal correction (QC-RLSC) and its machine-learning successor SERRF require strategically interleaved QC injections to learn temporal structure and reduce technical variance, thereby turning plate design and autosampler scheduling into quantitative variables rather than logistical afterthoughts [14, 15]. Because these corrections assume exchangeability of QC aliquots and stable extraction/injection behavior, sample handling parameters – including plant mass, comminution size, extraction temperature and duration, and post-extraction storage – must be standardized so that QC pools remain representative of the analytical sequence rather than forming an unintended secondary matrix. This direct connection between experimental design and normalization logic provides a natural transition to extraction choices aimed at maximizing chemical coverage while minimizing matrix-induced variance.

Extraction chemistry governs which constituents reach the detector, so solvent systems must reflect the polarity and stability spectrum of target classes and the matrix architecture of the herb or spice. Hydroalcoholic mixtures (methanol or ethanol in water) at mild acidity favor phenolic acids, flavonoid glycosides, and mid-polarity glycosides common to leaves, bark, and seeds, while ether-rich partitions enhance the recovery of lipophilic terpenoids and sterols by reducing co-extracted polar species that would otherwise intensify ion suppression [16]. Pressurized or accelerated liquid extraction, and ultrasound or microwave-assisted extraction, reduce both solvent consumption and processing time; however, temperature ceilings must be enforced to prevent de-glycosylation or oxidation of labile phenolics and sulfur-containing metabolites, as emphasized in recent natural-products metabolomics guidelines that outline class-specific stability windows [17–19]. The overarching principle is to balance broad chemical recovery with the constraints of chromatographic and electrospray ionization (ESI) compatibility, which, in turn, leads to cleanup strategies for matrix-effect control.

Matrix effects, defined as co-elution-driven alterations in ionization efficiency, can obscure true biological variation or generate artefactual differences in both untargeted and targeted analyses. Their evaluation must, therefore, be integrated into early method development rather than postponed to final validation. Post-column infusion maps suppression/enhancement across retention time, revealing "danger zones" that can be addressed by gradient redesign, stationary-phase choice, or selective cleanup, while post-extraction addition quantifies net suppression for targets and analogues [20]. QuEChERS-style dispersive SPE and polymeric mixed-mode cartridges, such as HLB or WAX, can remove chlorophylls, waxes, and polysaccharides without excessive analyte loss when screening parameters are tuned on a small design-of-experiments grid [21]. Because the topology of matrix effects depends on retention behavior, the separation strategy becomes the next pivotal determinant of spectral clarity and identification success.

Green extraction solvents, such as natural deep eutectic systems (NADES), are increasingly used to mobilize wall-bound polyphenols and enhance the solubility of highly polar acids, with the added benefit of LC compatibility when dilution and ion-pair suppression are controlled. However, viscosity and background ions demand attention to prevent chromatographic distortion in gradient starts [48]. For alkaloid-rich matrices, slightly basic extractions enhance recovery but risk co-extracting saponins and proteins, so a two-stage workflow – acidic defatting followed by basified back-extraction or SPE cleanup – often improves both coverage and electrospray stability [22]. In all cases, pooled QC preparation must mirror the dominant extraction and cleanup route so that drift-correction models learn the true analytical system rather than an artificially cleaner matrix, providing a logical bridge to chromatographic and ionization decisions that govern spectral purity and sensitivity.

Carryover control and stability assessments close the pre-analytical chain because viscous or resinous spice extracts can cause memory effects, while labile constituents (e.g., aldehydes, anthocyanins) may degrade during autosampler residence. Instrument-specific needle-wash compositions and stability-bracketing schemes must, therefore, be validated using representative analyte panels rather than generic surrogates [23]. Once sampling, extraction, and cleanup define what enters the column, chromatographic selectivity and mobile-phase chemistry become the primary levers for resolving co-elution and modulating adduct formation patterns.

3. Chromatographic selectivity and ionization: Controlling isomerism, adducts, and matrix effects

Orthogonal chromatographic strategies are the first defense against co-elution and chimeric MS/MS spectra because stationary-phase chemistry and gradient design redistribute chemical load and separate isomers that otherwise defeat spectral matching and *in silico* annotation. Reversed-phase C18 and phenyl-hexyl (or pentafluorophenyl) phases provide complementary π-interaction selectivity for flavonoids and phenylpropanoids, while HILIC (amine or amide phases) resolves small organic acids, glycosides, and amino-sugar conjugates that elute near void in RP, thereby increasing MS/MS purity and boosting the fraction of features eligible for high-confidence IDs [24]. For example, flavonoid C-glycosides that co-elute on C18 often become baseline-resolved on phenyl-hexyl columns because π-stacking interactions differentiate their aromatic cores. Where matrices are exceptionally congested, heart-cutting 2D-LC (HILIC to RP or RP to SFC) and supercritical fluid chromatography (SFC) with APCI/APPI interfaces further expand separation space for terpenoids, polyphenols, and sterols while maintaining throughput compatible with batch studies [25]. Because elution determines which co-extracts share droplet space in the ion source, post-column infusion mapping should accompany gradient development to identify suppression-prone windows and verify that redesigns shift sentinel analytes out of those zones [26]. This separation-centric development lays the foundation for ionization strategies that stabilize adduct chemistry and polarity response.

Mobile-phase additives are decisive for adduct patterns and ionization efficiency, particularly in negative mode, where small organic acids and lipids dominate many herbal matrices. Ammonium fluoride has emerged as a highly effective dopant that increases $[M–H]^-$ formation and signal without degrading chromatographic performance, often outperforming ammonium acetate/formate for acidic classes [21, 22, 27]. In positive mode, ammonium or sodium adduction can be exploited for glycosides and saponins but demands careful interpretation because adduct switching complicates library matching; here, library cleaning and adduct grouping steps in post-processing mitigate mis-annotation downstream [26, 28, 29]. Because additive choice reshapes both sensitivity and fragmentation chemistry, additive screens should be conducted in the presence of representative matrix extracts rather than neat standards to capture emergent suppression and adduct competition. These chemical considerations dictate how acquisition logic should allocate the instrument duty cycle to maximize interpretable spectra.

Interface and polarity also respond to class-specific chemistry: ESI generally favors phenolics, glycosides, and alkaloids, whereas APCI/APPI often improves sensitivity for less polar terpenoids and sterols without derivatization, thereby enabling LC-based quantification of compounds historically measured by GC alone [30, 31]. Fast polarity switching increases coverage but reduces ion sampling per polarity due to overhead; two-pass acquisitions or polarity-split extracts preserve MS/MS density for each class at the cost of doubled chromatographic time, a tradeoff best justified when matrix infusion maps show polarity-specific suppression windows [19, 20, 26, 32]. Together, these decisions about separation and ionization create the analytical conditions under which acquisition rules can generate structurally rich MS/MS data, guiding the transition to DDA, DIA, and IMS logic.

Temperature, pH, and injection solvent composition collectively complete the framework of separation-focused control. Column temperature alters chromatographic selectivity, apparent pKa values, and solvent viscosity, thereby modulating retention behavior [33]. Injection-solvent strength relative to the initial gradient directly influences peak shape and the frequency of MS/MS triggers, particularly for early-eluting polar metabolites. Shallow initial gradients and injection solvents that match the starting mobile phase minimize breakthrough and peak broadening, which would otherwise compromise MS/MS quality [34]. When integrated with online mapping of ion suppression and adduct formation, these chromatographic safeguards improve the efficiency and scientific value of downstream acquisition-logic optimization. With co-elution and ionization chemistry managed, acquisition strategies must be optimized to capture high-quality MS/MS at scale without sampling bias.

4. Acquisition logic: DDA, DIA, IMS coupling, and targeted PRM/MRM

Acquisition rules regulate how effectively an instrument converts chromatographic peaks into structurally informative MS/MS spectra. Controlled studies now quantify how DDA parameters influence identification yield and the structure of molecular networks in complex matrices [35, 36]. On Orbitrap-class instruments, dynamic exclusion timing, isolation-window width, normalized collision energy (including stepped NCE), and AGC or ion-target levels determine which precursor ions are fragmented and how frequently they are selected. Optimizing these settings using proxy indicators such as library-match rate, reduction of singleton nodes, and appropriate MS/MS placement improves spectral diversity and enhances downstream annotation in herb-rich datasets [35]. Intelligent DDA strategies, such as AcquireX, iteratively remove background ions and previously sampled features to prioritize fragmentation of low-abundance precursors; this often requires multiple reinjections but substantially increases spectral coverage without excessive runtime [37]. However, because classical DDA is inherently intensity-biased, a natural transition is to complement it with DIA, which overcomes undersampling in congested chromatographic regions.

DIA fragments all ions within sliding m/z windows and excels in reproducibility and quantitative stability but requires careful window design and robust deconvolution. Introducing DIA after DDA in the workflow provides continuity: DDA maximizes diversity, while DIA maximizes completeness. Coupling DIA with trapped ion mobility (e.g., dia-PASEF) improves precursor selectivity via drift separation and increases duty cycle utilization, now widely demonstrated in complex small-molecule matrices [12, 38]. Orthogonal ion-mobility technologies such as FAIMS and TIMS reduce chimericity and provide CCS constraints that filter candidate structures, with recent reviews documenting performance gains in food and plant metabolomics [39, 40]. To maintain inter-batch comparability, IMS should be calibrated against vendor or community CCS libraries, particularly when CCS-assisted library searching or SIRIUS-based class constraints are used downstream [10, 29]. This progression from DDA to DIA to IMS reflects an increasingly structured approach to maximizing spectral clarity, coverage, and confidence.

Targeted modes close the discovery-to-decision loop by delivering confirmation and quantification. Parallel reaction monitoring (PRM) on high-resolution instruments captures full product-ion spectra per transition for scheduled analytes, enabling post-hoc interference checks and ion-ratio confirmation, while triple-quadrupole MRM maximizes throughput and sensitivity for large marker panels in quality control and surveillance. Transition or product-ion selection should be informed by prior untargeted MS/MS and repository-scale searches (e.g., via MassQL) to ensure robust diagnostic fragments across laboratories and additive systems [12]. Since acquisition only creates value when processed transparently, the next section addresses post-acquisition pipelines that transform raw files into interpretable chemical knowledge.

5. Post-processing and data curation: From features to interpretable entities

Open and reproducible pipelines facilitate the standardized conversion of vendor raw files into aligned and annotated feature tables. These workflows encapsulate all essential preprocessing steps, including peak detection, deconvolution, feature grouping, retention-time alignment, gap filling, and spectral export [29, 41]. MZmine 3 integrates isotope/adduct logic and QC utilities and supports IMS data, while MS-DIAL 5 simultaneously processes DDA and DIA data and now extends to MS imaging, thereby allowing contiguous analyses of herbal extracts across acquisition modes within a single framework [7, 41]. Feature-based molecular networking in GNPS and allied graph methods then contextualizes spectra by structural relatedness, improving interpretability and supporting putative class assignments even when exact library matches are absent [13, 35]. Because library noise and inconsistent metadata degrade matching and machine-learning training, matchms supplies a robust, reproducible library-cleaning pipeline that harmonizes metadata, removes artifacts, and filters peaks to raise the true-positive rate of spectral search and stabilize analog discovery workflows. These curated outputs become the substrate for multi-evidence identification.

Multi-evidence annotation ladders information from empirical matches to *in silico* reasoning. SIRIUS/CSI:FingerID combines isotope-pattern analysis with fragmentation-tree scoring to generate molecular-formula and structure candidates, while CANOPUS adds class-level labels that remain useful when exact structures are absent. Current documentation details how library hits are integrated as annotations without unduly biasing ranking, thereby maintaining transparency [10, 42]. Retention-time prediction and transfer, through direct mapping (PredRet) or learned models (RT-Transformer) and post-projection calibration, filters candidates by penalizing implausible elution orders under a given chromatographic mode, with recent work demonstrating broad model transferability across LC systems [11, 43]. Ion-mobility CCS adds an orthogonal physicochemical descriptor that sharply reduces false positives in classes known to produce degenerate MS/MS, an increasingly common step in food-omics pipelines for herbs and spices [44, 45]. Because identification sets the ceiling for inference, the next section introduces learned spectral similarity and analog search tools that extend structural reach beyond exact library matches. Beyond these rule-based and physics-guided methods, learned spectral-similarity models introduce an additional layer of inference capable of capturing relationships that traditional matching strategies may overlook.

Learned spectral-similarity approaches, such as MS2DeepScore and integrated analog-search tools, like MS2Query refine the recovery of structurally meaningful neighbors for unknown metabolites. These tools embed MS/MS spectra into learned representational spaces in which distance correlates with structural similarity, thereby increasing the number of detected features that can be confidently associated with plausible scaffolds or chemical classes [8, 9]. When these spectral embeddings are incorporated into molecular-networking workflows, the resulting graphs form coherent communities of related metabolites rather than disconnected m/z features. This network-level organization enhances biomarker stability and deepens the mechanistic interpretation of complex herbal metabolite fingerprints. At the final stage before statistical analysis and quantitative modeling, QC-anchored normalization mitigates technical drift so that observed effect sizes, class separations, and calibration trends reflect true chemical differences rather than run-order artifacts. With structural information consolidated and drift minimized, the identification framework aligns directly with normalization procedures to ensure that downstream statistical inference remains robust and chemically meaningful.

6. Molecular identification and annotation: A multi-evidence paradigm

Identification in botanical metabolomics requires converging lines of evidence because extensive isomerism, conjugation patterns, and adduct variability limit the reliability of any single annotation cue. Empirical spectral matching in databases such as HMDB and MoNA/MassBank provides the initial layer of support [46, 47]. *In silico* fragmentation approaches, including SIRIUS and CSI:FingerID, extend this by ranking candidate structures based on predicted fragmentation behavior [10]. Retention-time prediction and transfer tools (PredRet, RT Transformer, post-projection calibration), CANOPUS class-label assignments, and ion-mobility CCS agreement then constrain structural candidates to reach MSI Level 2 or Level 3 identifications with explicitly stated confidence [11, 42, 43]. When exact spectral matches are not available, analog-search methods and learned spectral-similarity models propagate class-level and scaffold-level hints. These embeddings cluster unknowns with chemically recognizable families, thereby improving interpretability and guiding targeted MS/MS acquisition or authentic-standard validation. The systematic use of these approaches has increased annotation coverage in herb and spice datasets and shortened the path from unknown features to chemically meaningful entities.

Retention-time and CCS constraints warrant special emphasis because they are orthogonal to m/z and fragmentation behavior. These orthogonal parameters reduce false positives without overfitting annotations to instrument-specific spectral idiosyncrasies [48]. PredRet provides direct RT transfer when chromatographic geometry is comparable across systems, whereas transformer-based models generalize across broader chromatographic regimes. Recent evidence shows that post-projection calibration can harmonize predicted RT values to local systems with only modest training requirements [49]. Because these constraints are valuable only when recorded and shared, standardized reporting of RT and CCS accompanied by confidence intervals should be included in spectral exports and supplementary materials.

Once identification scaffolding is established, normalization becomes the critical gatekeeper that determines whether observed metabolic differences withstand the noise inherent to large-scale acquisitions. QC-anchored normalization transforms raw intensities into stabilized measures by modeling time-dependent drift and inter-feature covariation, ensuring that biological variation dominates residual technical noise in multivariate analyses and classification. SERRF learns feature-wise systematic error from QC injections and often outperforms classical LOESS in large cohorts, whereas QC-RLSC remains a transparent and robust default method when QC density is sufficient [14, 15]. Because normalization interacts directly with feature filtering and missing-value handling, imputation must follow censoring-consistent rules, such as left-censoring at the LOQ, rather than generic mean or median replacements that distort covariance structure. With identification and normalization secured, quantification algorithms can translate relative differences into decision-grade measurements. As these representational tools enrich structural context, normalization procedures ensure that the resulting chemical assignments translate reliably into quantitative and statistical analyses.

7. Normalization, drift correction, and QA/QC: Protecting inference over long sequences

Normalization anchored in pooled QC injections protects the integrity of statistical inference by explicitly modeling run-order drift and batch discontinuities. This approach lowers within-class %RSD and preserves true biological contrasts across the extended multi-day acquisitions that are typical of herb and spice metabolomics studies [14]. Such stabilization ensures that downstream analytical patterns reflect chemistry rather than temporal artifacts. QC-RLSC corrects signal variation by fitting a local polynomial function to each feature's intensity profile over time, allowing smooth modeling of drift. In contrast, SERRF applies random forest regression to learn feature-specific systematic error directly from QC injections, enabling a more flexible representation of complex drift patterns. Studies in large lipidomics cohorts demonstrate that SERRF commonly enhances classification performance and reduces technical variance beyond what QC-RLSC achieves, with improvement scaling to QC injection density and feature abundance [15]. This evidence highlights the value of nonlinear learning approaches when dealing with heterogeneous herbal matrices.

Normalization is necessary but insufficient when matrix effects dominate, so matrix-matched calibration, isotopologue internal standards for high-value targets, and post-column infusion mapping are complementary controls that tighten quantitative comparability [26]. Infusion maps, in particular, reveal suppression "valleys" that correlate with co-eluting matrix clusters and guide gradient and cleanup redesign, while post-extraction addition benchmarks net suppression/enhancement for sentinel analytes used to monitor batch health. Because complex herbal matrices may lack clean blanks, standard-addition calibration becomes a practical tool for targeted panels despite increased sample use and analyst effort, and should be reported with slope confidence intervals and matrix-specific recoveries. These method-centric controls reduce normalization burden and produce intensities more amenable to stable statistical modeling.

Feature curation and missing-value handling complete the QA/QC loop. Filter thresholds should be set with respect to detection frequency in QCs and biological replicates to avoid survivorship bias that inflates separation in PCA and PLS-DA; batch-wise and study-wise filters may be required when acquisition spans instrument maintenance [50]. For values censored below LOQ, left-censoring-aware imputation or replacement by a fraction of LOQ preserves distributional structure better than arbitrary small constants – an approach consistent with ICH expectations for quantitative reporting [3]. When combined with repository-scale reanalysis to verify marker stability across studies, these QA/QC practices produce feature tables and spectral evidence that stand up to regulatory and peer scrutiny, setting the stage for targeted quantification and method validation.

8. Quantification: From untargeted relative abundance to targeted decision-grade measures

Quantification strategies in herbal metabolomics follow a tiered structure because discovery-oriented studies, specification-grade identity and purity testing, and risk-driven surveillance impose fundamentally different requirements for calibration rigor, selectivity, and interference control. Discovery workflows aim for robust relative changes, whereas specification testing requires numerical accuracy and strict analytical specificity. This contrast defines the depth of validation and the complexity of the calibration model. Multi-analyte MRM panels on triple-quadrupole instruments remain the highest-throughput option for quantifying curated markers and detecting adulterants in herbal matrices, and their selectivity and linearity continue to support regulatory and industrial applications. PRM on Orbitrap-class instruments complements this approach by acquiring full product-ion spectra for each scheduled transition, which enables rigorous ion-ratio checks and post-hoc interference assessment without major sensitivity loss across many analyte classes [51]. These capabilities increase confidence when working with chemically complex plant matrices.

Because herbal matrices often lack clean blanks and authentic isotopologues for all targets, hybrid calibration approaches, including matrix-matched multi-point curves for critical analytes, internal standard single-point normalization for broader panels, and standard addition for stubborn matrices, are pragmatic compromises that must be reported with clarity about sources of uncertainty. Matrix-effect benchmarking and mitigation are inseparable from quantification in complex botanicals. Post-column infusion provides a chromatographic heatmap of suppression/enhancement that guides gradient shape, washouts, and valve timing, while post-extraction addition quantifies residual matrix influence under finalized gradients.

Selectivity safeguards ensure that reported concentrations remain valid. Scheduled PRM/MRM methods that monitor at least two product ions per analyte and verify ion ratios within preset tolerances effectively reduce false positives. High-resolution PRM adds chromatographic and exact-mass specificity, which becomes decisive when plant matrices contain near-isobaric isomers or co eluting species [3]. For multi-component specifications in herbal formulations, reporting should include the rationale for each marker (bioactivity, chemotaxonomy, safety), acceptance ranges justified by repository-scale evidence where available, and cross-checks with orthogonal stimuli-response endpoints when claims invoke pharmacological mechanisms [3, 52]. This explicit alignment of markers with functional endpoints integrates quantitative LC–MS/MS with systems-level interpretation and prepares assays for regulatory review or surveillance. Finally, method lifecycle thinking ties quantification back to discovery by using repository mining to check the uniqueness and robustness of diagnostic fragments and neutral-loss patterns across studies and instruments, and by depositing raw files and metadata in MetaboLights to allow independent reanalysis [12, 53]. Given the rapid evolution of supply chains and herbal formulations, this cycle – from fingerprint to marker, from validated assay to repository confirmation – ensures that quantitative results remain credible and actionable for clinicians, regulators, industry, and researchers.

9. Conclusion(s)

LC–MS/MS now offers a streamlined and standards-driven framework for phytochemical characterization by uniting selective extraction, orthogonal chromatographic separation, optimized ionization, and adaptive acquisition into a coherent workflow. These integrated advances enable a shift from narrow marker assays to comprehensive chemical fingerprints that simultaneously support authentication, quality evaluation, and mechanistic insight within a single analytical campaign. The robustness of these outcomes relies on transparent identification criteria, multi-evidence annotation strategies, and QC-anchored normalization embedded within open and interoperable processing pipelines. As spectral repositories expand and machine-learning tools mature, LC–MS/MS is poised to advance phytochemical analysis into a fully cumulative and regulation-ready discipline, reinforcing its central role in natural-product chemistry, functional-food innovation, and botanical therapeutics.

Acronyms and abbreviations

LC-MS/MS	Liquid chromatography–tandem mass spectrometry
HPLC	High-performance liquid chromatography
DDA	Data-dependent acquisition
DIA	Data-independent acquisition
PRM	Parallel reaction monitoring
MRM	Multiple reaction monitoring
HILIC	Hydrophilic interaction liquid chromatography
SFC	Supercritical fluid chromatography
ESI	Electrospray ionization
APCI	Atmospheric pressure chemical ionization
APPI	Atmospheric pressure photoionization
CID	Collision-induced dissociation
HCD	Higher-energy collisional dissociation
NCE	Normalized collision energy
m/z	Mass-to-charge ratio
CCS	Collision cross section
IMS	Ion mobility spectrometry
RT	Retention time,
QC	Quality control
QC-RLSC	Quality-control-based robust LOESS signal correction
SERRF	Systematic error removal using random forest
FBMN	Feature-based molecular networking
GNPS	Global Natural Products Social Molecular Networking
CSI:FingerID	Fragmentation tree–based structure prediction
CANOPUS	Structural class prediction
SIRIUS	Molecular structure prediction platform
MSL	Mass spectral library
NADES	Natural deep eutectic solvents
LOQ	Limit of quantification
PCA	Principal component analysis
PLS-DA	Partial least squares discriminant analysis

Conflict of Interest

The authors declare no conflict of interest.

Acknowledgments

The author gratefully acknowledges the use of the Gemini 2.5 Pro model (Google) to assist with comprehensive language editing and structural refinement of the manuscript.

Author details

Yodying Yingchutrakul[1] and Sucheewin Krobthong[2*]

1 National Center for Genetic Engineering and Biotechnology, NSTDA, Pathum Thani, Thailand

2 Center of Excellence in Natural Products Chemistry, Department of Chemistry, Faculty of Science, Chulalongkorn University, Bangkok, Thailand

*Address all correspondence to: sucheewin.k@chula.ac.th

References

[1] Altemimi A, Lakhssassi N, Baharlouei A, Watson DG, Lightfoot DA. Phytochemicals: Extraction, isolation, and identification of bioactive compounds from plant extracts. Plants (Basel, Switzerland). 2017;**6**. DOI: 10.3390/plants6040042

[2] Xue J-Y, Liu S-J, Wang J, Huang X-C, Xu Z-C, Fang -X-X, Li Z, Van de Peer Y. Multi-omics roadmap to plant-derived medicines. Engineering. 2025. DOI: 10.1016/j.eng.2025.10.019

[3] Li S, Siddiqa A, Thapa M, Chi Y, Zheng S. Trackable and scalable LC-MS metabolomics data processing using asari. Nature Communications. 2023;**14**:4113. DOI: 10.1038/s41467-023-39889-1

[4] Bhambhani S, Kondhare KR, Giri AP. Diversity in chemical structures and biological properties of plant alkaloids. Molecules. 2021;**26**:3374. DOI: 10.3390/molecules26113374

[5] Elbouzidi A, Haddou M, Baraich A, Taibi M, El Hachlafi N, Pareek A, Mesnard F, Addi M. Biochemical insights into specialized plant metabolites: Advancing cosmeceutical applications for skin benefits. Journal of Agriculture and Food Research. 2025;**19**:101651. DOI: 10.1016/j.jafr.2025.101651

[6] Schmid R, Heuckeroth S, Korf A, Smirnov A, Myers O, Dyrlund TS, Bushuiev R, Murray KJ, Hoffmann N, Lu M, et al. Integrative analysis of multimodal mass spectrometry data in MZmine 3. Nature Biotechnology. 2023;**41**:447–449

[7] Tsugawa H, Cajka T, Kind T, Ma Y, Higgins B, Ikeda K, Kanazawa M, VanderGheynst J, Fiehn O, Arita M. MS-DIAL: Data-independent MS/MS deconvolution for comprehensive metabolome analysis. Nature Methods. 2015;**12**:523–526. DOI: 10.1038/nmeth.3393

[8] de Jonge NF, Louwen JJR, Chekmeneva E, Camuzeaux S, Vermeir FJ, Jansen RS, Huber F, van der Hooft JJJ. MS2Query: Reliable and scalable MS(2) mass spectra-based analogue search. Nature Communications. 2023;**14**:1752. DOI: 10.1038/s41467-023-37446-4

[9] Huber F, van der Burg S, van der Hooft JJJ, Ridder L. MS2DeepScore: A novel deep learning similarity measure to compare tandem mass spectra. Journal of Cheminformatics. 2021;**13**:84. DOI: 10.1186/s13321-021-00558-4

[10] Dührkop K, Fleischauer M, Ludwig M, Aksenov AA, Melnik AV, Meusel M, Dorrestein PC, Rousu J, Böcker S. SIRIUS 4: A rapid tool for turning tandem mass spectra into metabolite structure information. Nature Methods. 2019;**16**:299–302. DOI: 10.1038/s41592-019-0344-8

[11] Stanstrup J, Neumann S, Vrhovšek U. PredRet: Prediction of retention time by direct mapping between multiple chromatographic systems. Analytical Chemistry. 2015;**87**:9421–9428. DOI: 10.1021/acs.analchem.5b02287

[12] Damiani T, Jarmusch AK, Aron AT, Petras D, Phelan VV, Zhao HN, Bittremieux W, Acharya DD, Ahmed MMA, Bauermeister A, et al. A universal

language for finding mass spectrometry data patterns. Nature Methods. 2025;**22**:1247–1254

[13] El Abiead Y, Strobel M, Payne T, Fahy E, O'Donovan C, Subramamiam S, Vizcaino JA, Yurekten O, Deleray V, Zuffa S, et al. Enabling pan-repository reanalysis for big data science of public metabolomics data. Nature Communications. 2025;**16**:4838

[14] Dunn WB, Broadhurst D, Begley P, Zelena E, Francis-mcintyre S, Anderson N, Brown M, Knowles JD, Halsall A, Haselden JN, et al. Procedures for large-scale metabolic profiling of serum and plasma using gas chromatography and liquid chromatography coupled to mass spectrometry. Nature Protocols. 2011;**6**:1060–1083

[15] Fan S, Kind T, Cajka T, Hazen SL, Tang WHW, Kaddurah-Daouk R, Irvin MR, Arnett DK, Barupal DK, Fiehn O. Systematic error removal using random forest for normalizing large-scale untargeted lipidomics data. Analytical Chemistry. 2019;**91**:3590–3596. DOI: 10.1021/acs.analchem.8b05592

[16] Gil-Martin E, Forbes-Hernandez T, Romero A, Cianciosi D, Giampieri F, Battino M. Influence of the extraction method on the recovery of bioactive phenolic compounds from food industry by-products. Food Chemistry. 2022;**378**:131918. DOI: 10.1016/j.foodchem.2021.131918

[17] Melikoglu M. Microwave-assisted extraction: Recent advances in optimization, synergistic approaches, and applications for green chemistry. Sustainable Chemistry for Climate Action. 2025;**7**:100122. DOI: 10.1016/j.scca.2025.100122

[18] Alves Filho EG, Lima M, Silva L, Ribeiro P, Tiwari BK, Fernandes FN, Brito ES. Green ultrasound-assisted extraction of bioactive compounds from button mushrooms, potatoes, and onion peels. ACS Food Science & Technology. 2021;**1**:1274–1284. DOI: 10.1021/acsfoodscitech.1c00153

[19] Nothias LF, Nothias-Esposito M, da Silva R, Wang M, Protsyuk I, Zhang Z, Sarvepalli A, Leyssen P, Touboul D, Costa J, et al. Bioactivity-based molecular networking for the discovery of drug leads in natural product bioassay-guided fractionation. Journal of Natural Products. 2018;**81**:758–767

[20] Taylor PJ. Matrix effects: The Achilles heel of quantitative high-performance liquid chromatography-electrospray-tandem mass spectrometry. Clinical Biochemistry. 2005;**38**:328–334. DOI: 10.1016/j.clinbiochem.2004.11.007

[21] Veiga-del-baño JM, Andreo-Martínez P, Pérez-Lucas G, Navarro S. Overview of the evolution and trends of the QuEChERS sample preparation procedure. Reviews of Environmental Contamination and Toxicology. 2024;**262**:22. DOI: 10.1007/s44169-024-00073-1

[22] Reyes-Garces N, Gionfriddo E, Gomez-Rios GA, Alam MN, Boyaci E, Bojko B, Singh V, Grandy J, Pawliszyn J. Advances in solid phase microextraction and perspective on future directions. Analytical Chemistry. 2018;**90**:302–360. DOI: 10.1021/acs.analchem.7b04502

[23] Cortese M, Gigliobianco MR, Magnoni F, Censi R, Di Martino PD. Compensate for or minimize matrix effects? Strategies for overcoming matrix effects in liquid

chromatography-mass spectrometry technique: A tutorial review. Molecules. 2020;**25**:3047. DOI: 10.3390/molecules25133047

[24] Hemmer S, Manier SK, Wagmann L, Meyer MR. Comparison of reversed-phase, hydrophilic interaction, and porous graphitic carbon chromatography columns for an untargeted toxicometabolomics study in pooled human liver microsomes, rat urine, and rat plasma. Metabolomics. 2024;**20**:49. DOI: 10.1007/s11306-024-02115-0

[25] Taglioni E, Capriotti AL, Cerrato A, Lagana A, Montone CM, Piovesana S, Cavaliere C. Recent applications of supercritical fluid chromatography in the analysis of polyphenols. Journal of Chromatography A. 2025;**1758**:466198. DOI: 10.1016/j.chroma.2025.466198

[26] Stahnke H, Reemtsma T, Alder L. Compensation of matrix effects by postcolumn infusion of a monitor substance in multiresidue analysis with LC-MS/MS. Analytical Chemistry. 2009;**81**:2185–2192. DOI: 10.1021/ac802362s

[27] McFadden JR, Ames DM. Assessment of ammonium fluoride as a mobile phase additive for sensitivity gains in electrospray ionization. Analytical Science Advances. 2023;**4**:347–354. DOI: 10.1002/ansa.202300031

[28] Liu S, Cui M, Liu Z, Song F, Mo W. Structural analysis of saponins from medicinal herbs using electrospray ionization tandem mass spectrometry. Journal of the American Society for Mass Spectrometry. 2004;**15**:133–141. DOI: 10.1016/j.jasms.2003.09.013

[29] de Jonge NF, Hecht H, Strobel M, Wang M, van der Hooft JJJ, Huber F. Reproducible MS/MS library cleaning pipeline in matchms. Journal of Cheminformatics. 2024;**16**:88. DOI: 10.1186/s13321-024-00878-1

[30] Kumar BR. Application of HPLC and ESI-MS techniques in the analysis of phenolic acids and flavonoids from green leafy vegetables (GLVs). Journal of Pharmaceutical Analysis. 2017;7:349–364. DOI: 10.1016/j.jpha.2017.06.005

[31] Cinquepalmi V, Losito I, Castellaneta A, Calvano CD, Cataldi TRI. APCI-Multistage mass spectrometry following liquid chromatography for selected 4-desmethyl-sterols and their deuterium-labelled analogues unveils characteristic fragmentation routes for cholesterol and phytosterols identification. Rapid Communications in Mass Spectrometry. 2025;**39**:e10039. DOI: 10.1002/rcm.10039

[32] Fitz V, El Abiead Y, Berger D, Koellensperger G. Systematic investigation of LC miniaturization to increase sensitivity in wide-Target LC-MS-Based trace bioanalysis of small molecules. Frontiers in Molecular Biosciences. 2022;**9**:857505. DOI: 10.3389/fmolb.2022.857505

[33] Soós B, Horváth K, Kormány R. The Dynamic balances in HPLC: An unconventional application of temperature to optimize ionizable solute separations. Journal of Chromatography A. 2025;**1760**:466268. DOI: 10.1016/j.chroma.2025.466268

[34] Naidong W, Chen YL, Shou W, Jiang X. Importance of injection solution composition for LC-MS-MS methods. Journal of Pharmaceutical and

Biomedical Analysis. 2001;**26**:753–767. DOI: 10.1016/s0731-7085(01)00439-3

[35] Stincone P, Pakkir Shah AK, Schmid R, Graves LG, Lambidis SP, Torres RR, Xia SN, Minda V, Aron AT, Wang M, et al. Evaluation of data-dependent MS/MS Acquisition parameters for non-targeted metabolomics and molecular networking of environmental samples: Focus on the Q exactive platform. Analytical Chemistry. 2023;**95**: 12673–12682

[36] Frohlich K, Fahrner M, Brombacher E, Seredynska A, Maldacker M, Kreutz C, Schmidt A, Schilling O. Data-Independent acquisition: A milestone and prospect in clinical mass spectrometry-based proteomics. Molecular & Cellular Proteomics. 2024;**23**:100800. DOI: 10. 1016/j.mcpro.2024.100800

[37] Razafindrabenja LE, Suc L, Veran F, Verbaere A, de Sousa Dias AL, Meudec E, Galy N, Tixador D, Loisel C, Mouls L, Sommerer N. Combining AcquireX data acquisition and Feature-Based Molecular Networking approach to deeply explore oxidation markers of condensed tannins after depolymerization. Food Chemistry. 2025;**483**:144207. DOI: 10.1016/j. foodchem.2025.144207

[38] Demichev V, Szyrwiel L, Yu F, Teo GC, Rosenberger G, Niewienda A, Ludwig D, Decker J, Kaspar-Schoenefeld S, Lilley KS, et al. dia-PASEF data analysis using FragPipe and DIA-NN for deep proteomics of low sample amounts. Nature Communications. 2022;**13**:3944

[39] Swearingen KE, Moritz RL. High-field asymmetric waveform ion mobility spectrometry for mass spectrometry-based proteomics. Expert Review of Proteomics. 2012;**9**:505–517. DOI: 10.1586/epr.12.50

[40] Houel S, Abernathy R, Renganathan K, Meyer-Arendt K, Ahn NG, Old WM. Quantifying the impact of chimera MS/MS spectra on peptide identification in large-scale proteomics studies. Journal of Proteome Research. 2010;**9**:4152–4160. DOI: 10.1021/pr1003856

[41] Heuckeroth S, Damiani T, Smirnov A, Mokshyna O, Brungs C, Korf A, Smith JD, Stincone P, Dreolin N, Nothias LF, et al. Reproducible mass spectrometry data processing and compound annotation in MZmine 3. Nature Protocols. 2024;**19**:2597–2641

[42] Duhrkop K, Nothias LF, Fleischauer M, Reher R, Ludwig M, Hoffmann MA, Petras D, Gerwick WH, Rousu J, Dorrestein PC, Bocker S. Systematic classification of unknown metabolites using high-resolution fragmentation mass spectra. Nature Biotechnology. 2021;**39**:462–471. DOI: 10.1038/s41587-020-0740-8

[43] Xue J, Wang B, Ji H, Li W. RT-Transformer: Retention time prediction for metabolite annotation to assist in metabolite identification. Bioinformatics. 2024;**40**. DOI: 10.1093/ bioinformatics/btae084

[44] Wisanpitayakorn P, Sartyoungkul S, Kurilung A, Sirivatanauksorn Y, Visessanguan W, Sathirapongsasuti N, Khoomrung S. Accurate prediction of ion mobility collision cross-section using ion's polarizability and molecular mass with limited data. Journal of Chemical Information and Modeling. 2024;**64**:1533–1542. DOI: 10.1021/acs. jcim.3c01491

[45] Díaz-Galiano FJ, Murcia-Morales M, Cutillas V, Fernández-Alba AR. Ions on the move: The combination of ion mobility and food metabolomics. Trends in Food Science & Technology. 2024;**147**:104446. DOI: 10.1016/j.tifs.2024.104446

[46] Wishart DS, Feunang YD, Marcu A, Guo AC, Liang K, Vazquez-Fresno R, Sajed T, Johnson D, Li C, Karu N, et al. HMDB 4.0: The human metabolome database for 2018. Nucleic Acids Research. 2018;**46**:D608–D617

[47] Elapavalore A, Kondic T, Singh RR, Shoemaker BA, Thiessen PA, Zhang J, Bolton EE, Schymanski EL. Adding open spectral data to MassBank and PubChem using open source tools to support non-targeted exposomics of mixtures. Environmental Science: Processes & Impacts. 2023;**25**:1788–1801. DOI: 10.1039/d3em00181d

[48] Lenski M, Maallem S, Zarcone G, Garcon G, Lo-Guidice JM, Antherieu S, Allorge D. Prediction of a large-scale database of collision cross-section and retention time using machine learning to reduce false positive annotations in untargeted metabolomics. Metabolites. 2023;**13**. DOI: 10.3390/metabo13020282

[49] Rampler E, Abiead YE, Schoeny H, Rusz M, Hildebrand F, Fitz V, Koellensperger G. Recurrent topics in mass spectrometry-based metabolomics and Lipidomics—Standardization, coverage, and throughput. Analytical Chemistry. 2021;**93**:519–545. DOI: 10.1021/acs.analchem.0c04698

[50] Lu J, Kerns RT, Peddada SD, Bushel PR. Principal component analysis-based filtering improves detection for Affymetrix gene expression arrays. Nucleic Acids Research. 2011;**39**:e86. DOI: 10.1093/nar/gkr241

[51] Rauniyar N. Parallel reaction monitoring: A targeted experiment performed using high resolution and high mass accuracy mass spectrometry. International Journal of Molecular Sciences. 2015;**16**:28566–28581. DOI: 10.3390/ijms161226120

[52] Alum EU, Manjula VS, Uti DE, Echegu DA, Ugwu OP-C, Egba SI, Agu PC. Metabolomics-driven standardization of herbal medicine: Advances, applications, and sustainability considerations. Natural Product Communications. 2025;**20**:1934578X251367650. DOI: 10.1177/1934578X251367650

[53] Yurekten O, Payne T, Tejera N, Amaladoss FX, Martin C, Williams M, O'Donovan C. MetaboLights: Open data repository for metabolomics. Nucleic Acids Research. 2024;**52**:D640–D646. DOI: 10.1093/nar/gkad1045

Chapter 5

Traditional Spices and Herbs in the Fight against Liver Cancer and Viral Hepatitis

Ali Irfan, Riaz Ullah, Tehreem Mushtaq, Huma Hameed, Yousef A. Bin Jardan, Shazia Perveen, Mahwish Arshad, Sadaf Ahmad, Bakar Bin Khatab Abbasi and Abdullah R. Alanzi

Abstract

The modern world, especially developing countries, is facing high morbidity and mortality rates due to liver cancer and viral hepatitis. The multidrug resistance in the standard viral hepatitis and cancer treatments has raised serious concerns and highlighted the urgency of novel therapeutic solutions. In this context, various herbal- and spice-based formulations, such as syrups, tablets, herbal teas, and natural supplements, are being explored both in clinical trials and in commercial markets. This chapter focuses on the role of traditional spices and herbs in preventing and managing liver cancer and viral hepatitis. Phytochemicals found in spices such as turmeric (curcumin), black seeds, ginger, and garlic demonstrate antiviral, antioxidant, anticancer, and anti-inflammatory potential. Herbs such as milk thistle, dandelion root, astragalus, coffee, barberry, plum, tamarind, and *Panax ginseng* are also used as adjunct therapies to protect the liver during hepatitis treatment. Bioactive compounds, including curcumin, triterpenoids, resveratrol, silibinin, tanshinone IIA, and piperine, have displayed anti-liver-cancer and antiviral therapeutic efficacy owing to their anti-fibrotic, anti-inflammatory, and antioxidant properties. Different phytoconstituents of spices and herbs, such as alkaloids, flavonoids, terpenoids, anthraquinones, phenols, phenylpropanoids, aromatics, and xanthones, exhibited hepatocellular carcinoma and hepatitis A, B, and C viruses. This chapter emphasizes on the recent spice- and herbs-based strategies for combating these diseases, focusing on their therapeutic potential, mechanism of action, and integration into modern medical approaches.

Keywords: hepatitis viruses, liver cancer, spice-derived compounds, herbal compounds, traditional medicine, phytochemicals, ginger and turmeric, silymarin

1. Introduction

In this modern era, liver diseases rank as the 11th most common cause of death, which is approximately 4% of deaths globally [1]. Hepatitis, also called viral hepatitis, is a fatal disease that causes complications due to which around 1–4 million deaths occur per year [2]. All the five types of hepatitis HAV, HBV, HCV, HDV, and HEV are caused by five viruses, namely hepatitis virus A, hepatitis virus B, hepatitis virus C, hepatitis virus D, and hepatitis virus E. Chronic hepatitis is life-threatening owing to liver cirrhosis and hepatocellular carcinoma (HCC) [3]. The hepatitis B virus is composed of DNA, while all other types of hepatitis viruses A, C, D, and E are composed of RNA [4–6]. Different hepatitis viruses spread through the following different ways as mentioned in **Table 1** [7–9].

2. Therapeutic role of spices and herbs in the treatment of hepatitis

Traditional medicines and herbal systems in India, Pakistan, Europe, and the USA, including Chinese herbal formulations, are the major sources of novel chemotherapeutic agents. The bioactive compounds derived from spices, herbs, and plants are administered to treat diseases in different dosage forms such as herbals teas, powders, tablets, capsules, and other herbal formulations [10, 11].

Different spices such as garlic (*Allium sativum*) significantly reduce weight by lowering total cholesterol, low density lipoprotein, and impart a significant effect on liver enzymes such as alanine aminotransferase (ALT) and aspartate aminotransferase, so garlic-based supplements are effective in treating chronic liver disease [12]. Another daily life common spice, ginger (*Zingiber officinale*), possessed hepatoprotective properties. Ginger supplements demonstrated anti-HCV results as

Hepatitis virus	Spreading Sources
Hepatitis A virus	HAV is transmitted from food and water contaminated with feces of individuals.
Hepatitis B virus	HBV can be transmitted vertically and through sexual contacts, such as vaginal secretions, semen, exposure to blood (e.g., drug abuse), or through injection and close person-to-person contact.
Hepatitis C virus	The major causes of HCV transmission are blood transfusions, contaminated healthcare utensils, injections, and also, sexual contact. HCV replication leads to inflammation and steatosis of the liver.
Hepatitis D virus	HDV is dependent on HBV surface antigen for its replication and transmitted like hepatitis HBV and HCV.
Hepatitis E virus	HEV can be transmitted via transfusion and zoonotic routes and also through contaminated food and water.

Table 1.
Spreading sources of hepatitis viruses.

well as showed reduction in nonalcoholic fatty liver disease due to reduction in inflammatory cytokines, ALT, insulin resistance index, γ-glutamyl transferase, and hepatic steatosis grade [13]. The compound IND02 (procyanidin type A) derived from cinnamon (*Cinnamomum verum*) inhibited HCV cell entry and provided a low-cost HCV hepatitis treatment [14].

Black seed (*Nigella sativa*) phytochemicals, such as catechin, rutin, and amento flavone, have the potential to inhibit the NS5B RdRp enzyme, which is essential for replication of the HCV [15]. Curcumin extracted from turmeric (*Curcuma longa*), a daily used spice, has the potential to inhibit the HCV replication by suppressing the signaling pathway Akt-SREBP-1, and curcumin also displayed the potential to inhibit PI3K-Akt, which indicated the anti-HCV therapeutic efficacy of turmeric [16]. Black peppers (*Piper nigrum*) and *Piper longum* are daily spices that have excellent chemotherapeutic significance in the medicinal field. These are composed mainly of piperine compound, which exhibited anti-HCV and anti-HBV properties [17, 18].

The mechanism of the HCV inhibition by the spices and herbs-derived compounds is given in **Figure 1.**

The herbs such as *Scytonema varium, Silybum marianum, Limonium sinense, Bupleurum kao, Camellia sinensis, Marrubium peregrinum L, and Antho cyanidin* contained anti-HCV compounds such as scytovirin (EC_{50}/CC_{50} = 17 nM/23.8 μM)

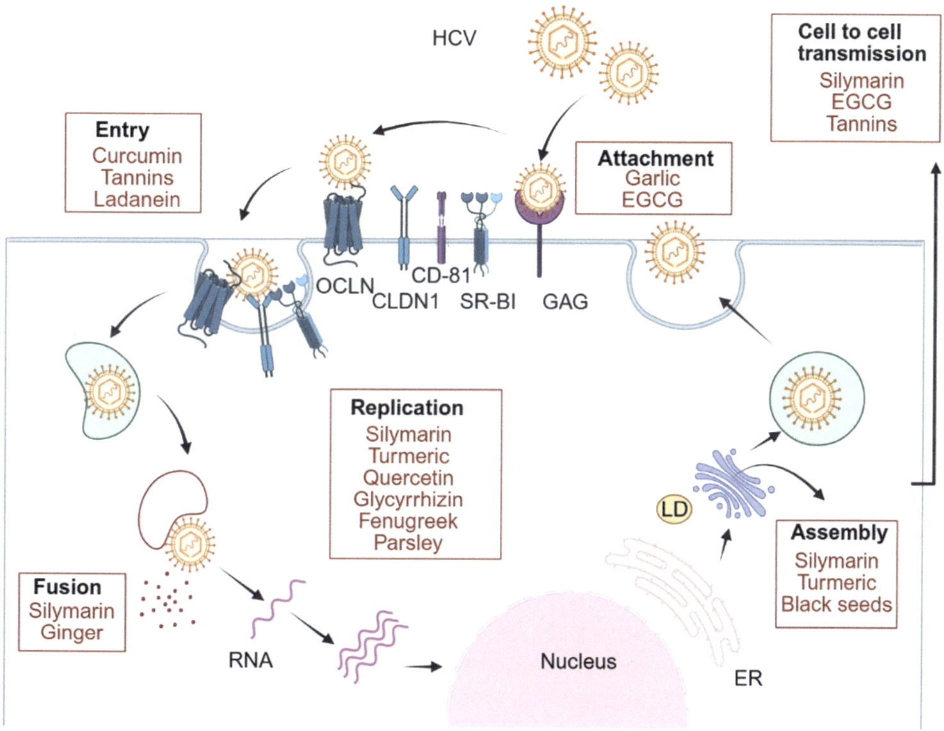

Figure 1.
HCV inhibition mechanism by spices- and herbs-derived compounds.

[19], silymarin (EC_{50} = 40–100 μM) [20–22], gallic acid (EC_{50}/CC_{50} = 24.31 ± 6.90 μM/ 346.59 ± 27.43 μM) [19], silibinin (EC_{50} = 40–100 μM) [23–25], saikosaponin B2 [19], EGCG [26–28], ladanein, delphinidin [29], and ladanein-BJ486K [30] and adopted GLUT2, RNA polymerase activity, NS5B, mevalonate, and entry inhibition mechanistic pathways. The compounds such as epigallocatechin-3-gallate [26–28], naringenin [31, 32], xanthone [33] hypophyllanthin [34], and cynaropicrin [35] of the following herbs *Camellia sinensis, grapefruit, phyllanthus niruri L., and cynara cardunculus* demonstrated entry inhibition metabolic pathway. Luteolin [36], 3-hydroxycaruilignan C [37], and plumbagin [38] extracted from *Petroselinum crispum, Swietenia macrophylla,* and *Plumbago indica L.* exhibited inhibition via shikimate, PI3K, Akt, mTOR, MAPK, NF-κB, and RNA replication pathways. Apigenin, xanthohumol, honokiol 5-O-methyl embelin, and apigenin extracted from *Petroselinum crispum, Humulus L, Magnolia grandiflora,* and *Embelia schimperi* (fruit) displayed inhibition activity β-oxidation, oxidative stress reduction, PI3K, Akt, NFκB, and STAT3 metabolic pathways [36, 39–41]. *Flavone* containing apigenin demonstrated anti-HCV activity via miR122 reduction [42]. Herbs such as *excoecaria agallocha L., alkaloid, ruta angustifolia, Maytenus ilicifolia,* and *Swietenia macrophylla* containing anti-HCV agents such as excoecariphenol D corilagin (EC_{50} = 13.5 μM), caffeine (EC_{50} = 0.726 Mm), chalepin (IC_{50} = 1.7 μg/Ml), APS, and 3-hydroxy caruilignan C (IC_{50} = 37.5 μM) exhibited replication inhibition. Pseudane IX [43] drawn out from the *Swietenia macrophylla* displayed anti-HCV activity through insulin resistance at EC_{50} = 1.4 μg/mL. Lipid metabolism was observed in quercetin extracted from *Embelia ribes* [20, 44]. Oleanolic acid extracted from exhibited NS5B polymerase inhibition at EC_{50} = 0.8–3.5 μg/mL [45]. The compounds piperine and coumestans were separated from *Piper longum L.*, and *Eclipta prostrata L.* showed anti-HBV and anti-HCV activity via inhibition of HBsAg and HBeAg [17, 18]. The compound glycycoumarin separated from the herb *Glycyrrhiza uralensis* displayed glucuronide action with EC_{50} = 4.6–8.8 μg/Ml [46]. The compound phyllanthin found in the herb *Phyllanthus niruri L.* exhibits anti-HCV activity through NS3/4A protease inhibition [47]. The herb *Maytenus ilicifolia* containing the compound induced apoptosis and autophagy modulation at IC_{50} = 2.3 μM [48]. The compound allicin [12, 49] separated from *Allium sativum L.* and α-mangostin [33] separated from *Garcinia mangostana* induced apoptosis and modulated anti-HCV activity. Phase I metabolism and epoxidation oxidation were observed in the compound limonin drawn from the herb *Ficus religiosa L* [50]. The compound 1-cinnamoyl-3,11 dihydroxy meliacarpin extracted from the herb *artocarpus* induced NF-κB modulation [51]. Glycyrrhizin extracted from the herb species *Glycyrrhiza uralensis* played an important role in anti-HCV activity through glycyrrhetinic acid conversion with IC_{50} = 16.5 μM [47, 52].

Rosmarinic acid [53] separated from the specie *lamiaceae* exhibited anti-HBV activity via inhibiting replication by binding to ε polymerase. *N*-acetylphenyl alanine, 6-hydroxyl-7-methoxyl-coumarin, iridoid lactone, magnatrol B, anislactone B, honokiol, isomagnolol, isocarpine, swertisin, and 1,5,8-trihydroxy-3-methoxyxanthone extracted from the herbs such as *P. crinitum, S. punicea, S. asper, L. henryi,* and *I. tectorum maxim* exhibited anti-HCV activity by HBsAg and HBeAg inhibition [54–56, 60, 57–59]. The compounds cichoric acid, crypto-chlorogenic acid, 2-methyl-6-phenyl-4*H*-pyran-4-one, 3-caffeoylquinic acid, *p*-hydroxybenzoic acid, 3,5-dicaffeoylquinic acid, 4,5-dicaffeoylquinic acid, 3,4-dicaffeoylquinic acid, and 3,4-dihydroxybenzoic acid extracted from *C. intybus, S. asper, A. capillaris, L. japonica, S. mussotii,* and *A. scoparia* exhibited anti-HBV activity by inhibition of HBsAg, HBeAg, and HBV-DNA [56, 60–66]. The compound rosmarinic acid separated from the herb *C. intybus* displayed anti-HBV activity via ε-Pol binding mechanism [67]. The compound gedunin drawn from the specie *azadirachta indica and A. Juss* inhibited NS3 RNA polymerase and NS3 protease [68]. The herb specie *shikuwasa limes* contains a compound 3,5,6,7,3',4'-hexamethoxyflavone that has been reported as an anti-HVB agent via inhibition of TNF-α [69]. The compound betulinic acid extracted from *scutellaria baicalensis* demonstrated anti-HBV by inhibiting the SOD2 enzyme [16–18, 70–72]. cccDNA inhibition was demonstrated by the compound norbellidifolin drawn from the herb *S. mussotii* and hyperoside, an herb derivative of *Abelmoschus manihot (L) medik* [59, 73, 74]. Wogonin, an herb derivative of *Scutellaria baicalensi,* displayed anti-HBV activity by replication inhibition mechanism [69, 75]. Allicin derived from *Allium sativum L.* showed anti-HBV activity by apoptosis induction [76]. Epigallocatechin-3-gallate derived from *Camellia sinensis (L.) Kuntze* revealed anti-HVB activity by inhibiting viral DNA and RNA synthesis [77]. Luteolin7-O–b–D -glucopyranoside, derived from the herb *Taraxacum mongolicum,* acts as anti-HBV agent by Akt signaling mechanism.

3. Potential of spices and herbs in the management of liver cancer

Liver cancer ranks among the top five cancers in the whole world in terms of mortality and is the third leading cause of cancer-related deaths globally [78, 79]. HCC is one of the most common cancers worldwide, and most current therapies are of limited efficacy. Natural compounds extracted from plants, herbs, and daily-used spices demonstrated promising potential in drug discovery and development of novel therapeutic agents for various diseases; notably over 60% of cancer treatment drugs are sourced from natural herbs, plant and spices, etc. [80–82]. Lycii fructus (wolfberry) decreased liver inflammation and fibrosis, and improved liver histology by reducing apoptosis [83]. Xanthorrhizol derived from *curcuma xanthorrhiza* inhibited the HepG2 cells via Bcl2 family members [84]. Green tea (*Camellia sinensis*) leaves contained a phenolic epigallocatechin-3-gallate (EGCG) compound which displayed promising potential against HCC and HepG2 cells through kinases signaling pathways [85]. An alkaloid berberine is derived from rhizome of coptidis which showed anti-proliferative HCC activity [86], and a phytoalexin called as resveratrol is obtained from red grapes exhibited anti-proliferative and anti-apoptosis properties [87]. DAS (diallyl sulfide) is a major constituent of garlic, which showed inhibitory anti-HCC activity through inhibition of TNF-α and IL-6 pathways [88]. In multidrug resistance (MDR) cancer therapy, a ginger compound 6-shogoal displayed promising anticancer potency against hepatoma cells by trigging ROS-mediated, caspase-dependent apoptosis [89].

A lot of compounds derived from plants, herbs, and spices have been studied for their anti-HCC potential via various metabolic pathways. Ursolic acid derived from *Gardenia jasminoides* [90, 91] has been reported as anti-HCC agent via oxidative stress reduction. A spice derivative curcumin derived from *Curcumina kwangsiensis* revealed anti-liver-cancer potential via inhibition of PCNA and CDC2 pathways [92]. The compound asparanin A extracted from the fungus specie *Asparagus officinalis L* exhibited inhibition of cyclin A, CDK1/4, and caspase 3/8/9, and played a significant role as anti-HCC agent [93]. The compound acetylshikonin derived from *Lithospermum erythrorhizon* [94] and methyl nonacosa-14E, 18E, 22E, and 26E-tetraenoate derived from the herb *amaranthus spinosus* displayed anti-HCC activity via downregulation of BCl-2 and BAX [95]. Chalcone flavokawain B extracted from the rhizomes of alpinia plants acted as anti-liver-cancer agent via PIK3/Akt metabolic pathway [96]. Andrographolide separated from the *andrographis paniculata* [97] plant and the compound boldine [98] derived from the plant *Peumus boldus* played an important role in anti-liver-cancer activity via apoptosis induction, anti-angiogenesis, and overexpression of BAX. The plant species *Dicranostigma leptopodum (maxim.) Fedde* containing compound isocorydine acted as anti-HCC agent through inhibition of p38 and ERK metabolic pathways [99]. Celastrol extracted from the plant *Tripterygium wilfordii* suppressed Akt activation and acted as anti-liver-cancer agent [100]. The compound furanodiene derived from the plant *curcuma wenyujin* exhibited anti-liver-cancer activity through MAPK pathway [101]. Astrakurkurone, a derivative of plant *astraeus hygrometricus,* demonstrated anti-HCC activity via upregulation of pro-apoptotic pathway [102]. The compound 2-methoxyjuglone derived from plant species *Juglans cathayensis* [103], apigenin separated from *alpinia officinarum* [104, 105], and citrus fruit derivative Naringenin [106] acted as anti-liver-cancer agents through caspase pathway. The compound resveratrol separated from the plant *Arachis hypogaea* [107] and boldine derived from *Peumus boldus* [98] induced apoptosis and pro-apoptosis of BAX to inhibit the HCC. The compound caffeic acid phenyl ester separated from *propolis* [108] and berberine derived from *Berberis vulgaris* revealed anti-liver-cancer potential through NF–kB metabolic pathway. The compound hispolon derived from *Phellinus linteus* revealed anti-HCC potential via inhibiting PI3K/Akt, ERK1/2 metabolic pathways [109]. The plant *Artemisia capillaris* contains a compound artemisinin, which has been reported as an anti-liver-cancer agent via PI3K/Akt and mTOR metabolic pathways [110]. Anti-cancer potential of the compound galangin derived from the herb *alpinia officinarum* was observed through PKC and ERK metabolic pathways [111]. The herb *Berberis vulgaris* contains a compound berberine that has been observed as anti-liver-cancer agent via NF–kB metabolic pathway [112]. Plumbagin, separated from the herb *Plumbago zeylanica,* acted as anti-HCC agent by binding to NF–KB and AP-1 receptors [113]. Betulinic acid derived from *betula pubescens betula* [114], boldine [98] extracted from the herb *Peumus boldus,* farnesol derived from *Vachellia farnesiana* [115], and apigenin derived from the plant *alpinia officinarum* [104, 105] exhibited anti-liver-cancer efficacy via apoptosis, over expression and pro-apoptotic BAX metabolic pathway. Platycodin D extracted from the herb *Platycodonis radix* exhibited anti-liver-cancer efficacy via MEK and ERK pathways [116]. *Allium sativum's* derivative, allicin, has been studied as anti-HCC agent working via p53 and PI3K/mTOR pathways [117]. The compound kaempferol separated from the plant *Brassica oleracea var* revealed anti-HCC efficacy through inhibition of MAPK and HIF-1 [118]. The compound theaflavins extracted from *Camellia sinensis* also exhibited anti-liver-cancer efficacy by blocking the STAT3 metabolic pathway [119]. The spice derivative, curcumin, derived from the *curcuma longa L.* displayed anti-liver-cancer potential by targeting the HepG2 cells

[120]. *Citrus limon* contains a compound, eriocitrin, which is known for anti-HCC efficacy by upregulation of p53, cyclin A, cyclin D3, and CDK6 [121]. Honokiol is a plant derivative of *Magnolia officinalis* acting as an anti-HCC by destabilizing the EGFR [122]. The plant *Camellia sinensis* contains a compound, EGCG, which is studied for its anti-liver-cancer activity via inhibition of PI3K/Akt metabolic pathways [123]. The compound capsaicin, a spice derivative of *capsicum annuum* [124], and the compound cantharidin extracted from *Lytta vesicatoria* [125] are known for their efficacy as anti-HCC agent via tumor growth inhibition mechanism [126]. The plant *scutellaria baicalensis* contains a compound wogonin that has been reported for its anti-HCC activity through inactivation of Akt signaling pathway [124]. The compound eriodictyol is derived from the plant *eriodictyon californicum* inhibiting HCC via upregulation of BAX and PARP [127]. The compound nitidine chloride extracted from the plant *Zanthoxylum nitidum* revealed cytotoxicity toward the Huh-7 cells [128]. The compound extracted from the *Pueraria lobata* acts as anti-HCC agent through regulation of MAPK metabolic pathway [129]. Apigenin separated from the plant *Apium genus* inhibits the growth of HepG2 and Huh-7 cell lines [130]. The compound tanshinone IIA derived from the plant *Salvia miltiorrhiza Bunge* acted as anti-HCC agent by targeting the ASGPR receptors [131]. The compound luteolin, a derivative of *reseda odorata,* modulated the caspase activation and revealed it efficacy as anti-HCC [132, 133]. Luteolin-7-O-glucoside derived from the *Cynara scolymus* exhibited anti-HCC potential by arresting the G2/M phases of cell [134]. The compound acetylshikonin extracted from *lithospermum erythrorhizon* revealed anti-liver-cancer efficacy by downregulation of BCl-2 and BAX [94]. Similarly, the compound eriodictyol extracted from *Eriodictyon californicum* exhibited anti-HCC activity by upregulation of BAX and PARP [135].

Stimulatory pathways are the activation of cellular defense mechanisms intrinsic to the cell that oppose tumor growth and viral persistence. These mechanisms are largely the induction of apoptosis, augmentation of metabolic regulation against cancer cells, and the activation of immune responses [136]. Moringa (*Moringa oleifera*) is rich in bioactive molecules such as quercetin, kaempferol, and isothiocyanates that induce upregulation of pro-apoptotic proteins such as BAX, BAK, PUMA, BIM, and NOXA. This induction induces mitochondrial outer membrane permeabilization, cytochrome c release, and caspase cascade activation, culminating in hepatoma cell programmed death [137]. Curcumin, the major polyphenol of turmeric (curcuma longa), targets cellular epigenetic regulators to affect metabolic pathways such as glycolysis and the Krebs cycle. Through glycolytic enzyme downregulation and promotion of oxidative phosphorylation, curcumin makes the environment hostile for the survival of cancer cells. Curcumin also disrupts BCL-2 family protein function, making cells apoptotic to sensitive [138]. Allicin from garlic and thymoquinone from black cumin (*Nigella sativa*) downregulate pro-survival proteins (MCL-1, BCL-2, BCL-W, BCL-XL) and block transcription factors controlling these proteins. This dual effect not only eliminates survival benefits from cancer cells but also triggers reactivation of apoptotic machinery. Also, thymoquinone has been shown to modulate NF-κB and Myc, the primary transcription factors in cancer cell proliferation, thus re-establishing normal control over the cell cycle [139].

Through these routes, the classic spices and herbs act to trigger the body's inherent tumor-suppressive function and are hence valuable as preventive and therapeutic agents against liver cancer and chronic viral hepatitis as indicated in **Figure 2**.

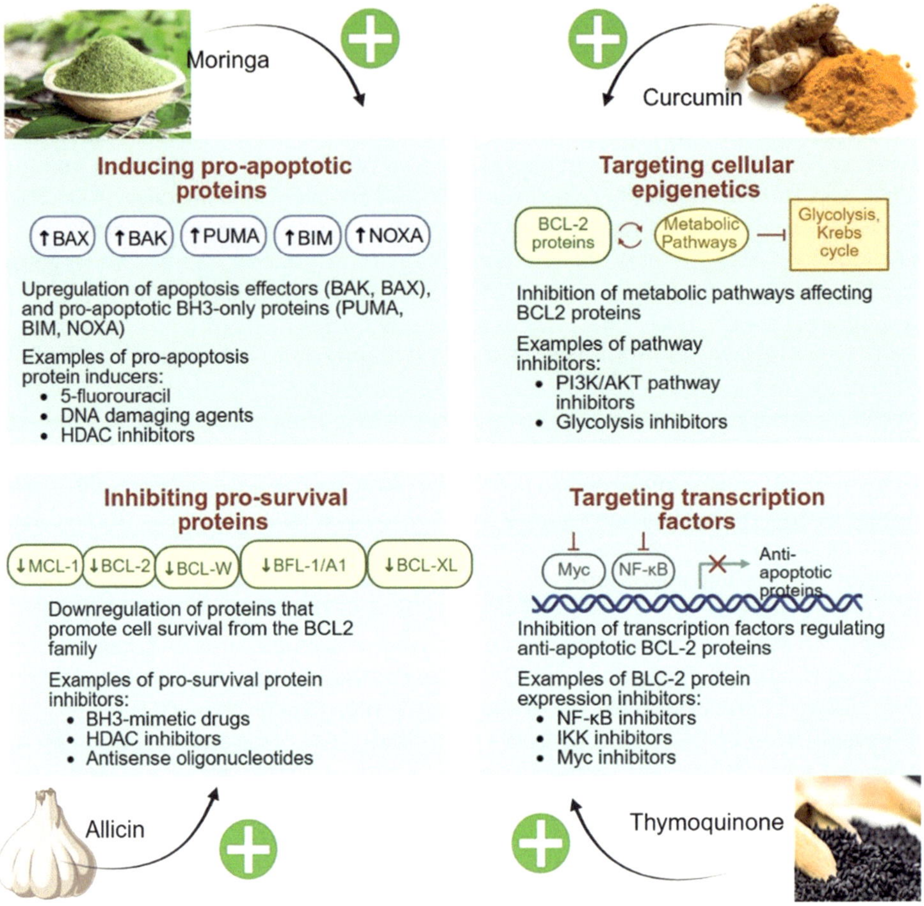

Figure 2.
Mechanistic pathways of HCC with inhibiting herbs and spices.

Figure 2 presents the stimulatory pathways by which specific traditional herbs and spices such as moringa, curcumin, allicin, and thymoquinone act to show anti-cancer action in liver cancer and viral hepatitis. Moringa induces apoptosis through upregulation of pro-apoptotic proteins (BAX, BAK, PUMA, BIM, NOXA), whereas curcumin acts on cellular epigenetics by blocking metabolic pathways affecting BCL-2 proteins. Allicin suppresses pro-survival BCL-2 family proteins and decreases cell survival signals. Thymoquinone disrupts the transcription factors NF-κB and Myc and inhibits expression of anti-apoptotic proteins. These bioactive molecules in combination promote programmed cell death and the inhibition of tumor development.

Inhibitory mechanisms aim at oncogenic and pro-inflammatory cascades of signals responsible for driving tumor growth, viral replication, and immune evasion. By inhibiting these pathways, conventional spices and herbs arrest disease progression at a number of molecular checkpoints [140]. Cinnamon polyphenols suppress PI3K/Akt signaling, a pathway commonly hyperactivated in HCC and

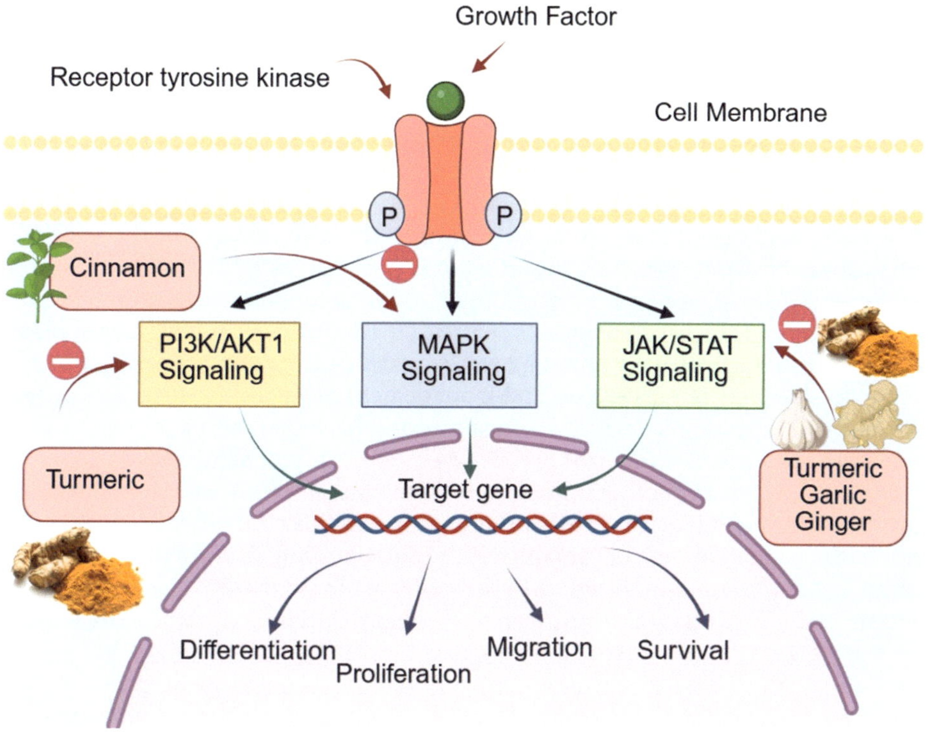

Figure 3.
HCC inhibitory mechanism by spices- and herbs-derived compounds.

chronic hepatitis, causing reduced proliferation, impaired glucose metabolism, and diminished metastatic potential [141]. Turmeric has a double inhibitory action – downregulating MAPK signaling to limit cell proliferation and inhibiting JAK/STAT signaling to hinder inflammation and anti-apoptotic gene transcription. It has a decisive role in limiting both tumor burden and chronic inflammatory injury in viral hepatitis [142]. Garlic and ginger synergize to suppress JAK/STAT signaling, lowering levels of inflammatory cytokines IL-6 and TNF-α, commonly elevated in HBV – and HCV-induced liver disease. This suppression of inflammatory signaling makes the environment less conducive to cancer progression and viral replication [143] as displayed in **Figure 3**.

Figure 3 illustrates inhibitory mechanisms by which classical spices and herbs such as cinnamon, turmeric, garlic, and ginger regulate crucial oncogenic signaling in viral hepatitis and liver cancer. Cinnamon suppresses the PI3K/Akt1 pathway, whereas turmeric represses MAPK and JAK/STAT pathways, effectively inhibiting tumor-promoting transcriptional activity. The synergistic interaction of turmeric, garlic, and ginger inhibits JAK/STAT-induced survival signals and downregulates cell proliferation, migration, differentiation, and survival genes. These inhibitory actions are responsible for anticancer growth prevention as well as viral pathogenesis prevention.

4. Conclusion

Liver cancer and viral hepatitis are major challenges and significant threats to the health of humans globally, particularly in developing countries. The developing countries are facing high mortality rates, which are compounded by the limitations in treatment facilities and conventional therapies and day by day, the growing threat of MDR. The major reasons for HCC are inborn metabolic disorders, oxidative stress, HCV, environmental toxins, chronic inflammation, HBV, some of drugs, etc. Accumulating evidence suggested that dietary spices, herbs, and plant-based products and their active compounds can inhibit liver cancer and replication of different hepatitis viruses. In this context, herbs, plants, and traditional spices have emerged as alternative therapeutic sources and valuable allies in the prevention and management of hepatitis and liver cancer. These natural products derived from spices and herbs can also act as adjunct therapies. The spices and herbs are rich in diverse phytochemicals such as resveratrol, curcumin, triterpenoids, piperine, and silibinin, and demonstrated anti-viral, anti-oxidant, and anti-inflammatory. In this context, traditional spices and herbs have emerged as valuable allies in prevention, management, and adjunct therapy. Rich in diverse phytochemicals, these natural agents exert potent antiviral, antioxidant, anticancer, and anti-inflammatory activities. Herbs such as barberry, milk thistle, panax ginseng, astragalus, dandelion, and traditional spices such as turmeric, ginger, black seeds, and garlic provide both experimental and clinical evidence that these herbs and spices have anti-HCC and anti-hepatitis potential by modulating key pathogenic pathways, inhibiting viral replication, reducing oxidative stress, and protecting hepatocytes. The other prominent botanicals, such as *Glycyrrhiza glabra*, wolfberry, broccoli sprouts, green tea, *Nigella sativ*a, red grapes, and coptidis rhizome, are bioactive compounds-based and capable of intercepting the progression of chronic hepatitis to HCC. The progression of hepatitis and liver cancer can be prevented by providing a shield of these natural products by integrating these natural products into nutrition and lifestyle to enhance therapeutic outcomes. In conclusion, combining traditional herbal research outcomes with modern biomedical research holds great promise to develop accessible, safe, effective, and efficient strategies to combat viral hepatitis and liver cancer.

5. Limitations and future perspectives

In defiance of the promising efficacy of herbs and spice derivatives against hepatitis viruses and liver cancer, there are still various pros. One of the substantial confrontations is the need for standardized formulations of herbs because differences in farming, garnering, separation, and preservation techniques lead to variation in the composition of bioactive compounds. In addition, studies based on *in vitro* and *in vivo* randomized clinical trials provided conclusions that are still meager. Drug-likeness problems such as fast metabolism, poor GI absorption, and lack of bioavailability limit the clinical use of biologically active compounds such as silibinin, resveratrol, and curcumin. Furthermore, herbal drug interaction and potential liver toxicity at high doses have highlighted the issues about long-term safety, especially when used with traditional therapies. In developing countries, there is insufficient use of herbal medicines and a lack of regulatory frameworks that restricts the extensive incorporation of innovative healthcare facilities.

From a broader point of view, ongoing research should validate clinical trials of herbs as a priority and efficiently design the herbal formulations and establish multicentric trials to gain dosage and long-term safety profiles. Implementation of advanced and innovative nanotechnology and drug delivery mechanisms has efficiently increased the bioactivity and efficacy of phytochemicals. The synergetic drug–herb and herb–herb interaction study can provide solutions for the multi-drug resistance issues. Furthermore, a broader and deeper understanding of signaling pathways and metabolic pathway modulation by herbal drugs is crucial for efficient application. There is a need to develop standard regulatory guidelines and quality control management to corroborate safety and reproducibility, and integration of personalized approaches such as phytotherapy with nutrigenomics and pharmacogenomics for individual treatments. In conclusion, implementation of preventive measures such as herbal drugs, dietary therapies, and lifestyle changes, along with cost-effective production and accessibility to resources, can help in the implementation of a global healthcare system and decrease the rate of viral hepatitis and liver cancer.

Acknowledgments

The authors gratefully acknowledge the Green Institute of Chemical, Biomedical, and Environmental Sciences (GICBES) for providing facilities and support for the preparation of this chapter.

Conflict of interest

The authors declare no conflict of interest.

Author details

Ali Irfan[1,2]*, Riaz Ullah[3], Tehreem Mushtaq[4], Huma Hameed[4], Yousef A. Bin Jardan[5]*, Shazia Perveen[6], Mahwish Arshad[7,8], Sadaf Ahmad[9], Bakar Bin Khatab Abbasi[10] and Abdullah R. Alanzi[3]*

1 Department of Chemistry, Government College University Faisalabad, Faisalabad, Pakistan

2 The Green Institute of Chemical, Biomedical and Environmental Sciences, Lahore, Pakistan

3 Medicinal, Aromatic and Poisonous Plants Research Center, College of Pharmacy, King Saud University, Riyadh, Saudi Arabia

4 Faculty of Pharmaceutical Sciences, University of Central Punjab, Lahore, Pakistan

5 Department of Pharmaceutics, College of Pharmacy, King Saud University, Riyadh, Saudi Arabia

6 Department of Zoology, The Women University Multan, Mattital Campus, Multan, Pakistan

7 Roy & Diana Vagelos Laboratories, Department of Chemistry, University of Pennsylvania, Philadelphia, PA, United States

8 Department of Medicine, Perelman School of Medicine, University of Pennsylvania, Philadelphia, PA, United States

9 Department of Physics, Chemistry and Pharmacy, University of Southern Denmark, Odense, Denmark

10 Department of Chemistry and Chemical Biology, Northeastern University, Boston, MA, United States

*Address all correspondence to: raialiirfan@gmail.com; ybinjardan@ksu.edu.sa; aralonazi@ksu.edu.sa

References

[1] Yin SY, Wei WC, Jian FY, Yang NS. Therapeutic applications of herbal medicines for cancer patients. Evidence-Based Complementary and Alternative Medicine. 2013;**2013** (1):302–426. DOI: 10.1155/2013/302426

[2] Zarrin A, Akhondi H. Viral hepatitis. In: StatPearls. Treasure Island (FL): StatPearls Publishing; 2021. Available from: https://www.ncbi.nlm.nih.gov/books/NBK556029/#

[3] González ME, González VM, Montaño MF, Medina GE, Mahadevan P, Villa C, Villa R. 2017. Genome-wide association analysis of body conformation traits in Mexican Holstein cattle using a mix of sampled and imputed SNP genotypes. Genetics and Molecular Research. **16**(2):160–235. DOI: 10.4238/gmr16029597

[4] AlMalki WH, Shahid I, Abdalla AN, Johargy AK, Ahmed M, Hassan S. Virological surveillance, molecular phylogeny, and evolutionary dynamics of hepatitis C virus subtypes 1a and 4a isolates in patients from Saudi Arabia. Saudi Journal of Biological Sciences. 2021;**28**(3):1664–1677. DOI: 10.1016/j.sjbs.2020.11.089

[5] Sinn DH, Cho EJ, Kim JH, Kim DY, Kim YJ, Choi MS. 2017. Current status and strategies for viral hepatitis control in Korea. Clinical and Molecular Hepatology. **23**(3):189–195. DOI:10.3350/cmh.2017.0033

[6] Shabanah W, Bukhari A, Alandijani A, Alyasi A, Youssef A-R. Prevalence of HBV and assessment of hepatitis B vaccine response among dental health care workers in dental teaching hospital, Umm Al-Qura university, Saudi Arabia. The Egyptian Journal of Immunology. 2019;**26**(2):11–17

[7] Ringelhan M, McKeating JA, Protzer U. Viral hepatitis and liver cancer. Philos. Soc R. London B, Biological Sciences. 2017;**372**(1732):201-274. DOI: 10.1098/rstb.2016.0274

[8] Zuckerman AJ. Hepatitis Viruses. In: Baron, S. Medical Microbiology. 4. Galveston (TX): University of Texas Medical Branch at Galveston; 1996. Available from: https://www.ncbi.nlm.nih.gov/books/NBK7864/

[9] Netter HJ, Barrios MH, Littlejohn M, Yuen LKW. Hepatitis delta virus (HDV) and delta-like agents: Insights into their origin. Frontiers in Microbiology. 2021;**12**:652–672. DOI: 10.3389/fmicb.2021.652962

[10] Balick MJ, Cox PA. Ethnobotanical research and traditional health care in developing countries. In: Bodeker G, Bhat KKS, Burley J, Vantomme P, editors. Medicinal Plants for Forest Conservation and Health Care. Rome, Italy: Food and Agriculture Organization of the United Nations; 1997. pp. 12–23

[11] Samuelsson G, Bohlin L. Drugs of Natural Origin: A Textbook of Pharmacognosy. 5th ed. Stockholm, Sweden: Swedish Pharmaceutical Press; 2004

[12] Xiaohui L, Jinqi L, Xiaofang X, Zhiqiang S, Renxiu N. 2023. Garlic supplementation for the treatment of chronic liver disease: A meta-analysis of randomized controlled trials. African Health Sciences. **23**(2):409–415. DOI: 10.4314/ahs.v23i2.47

[13] Rahimlou M, Yari Z, Hekmatdoost A, Alavian SM, Keshavarz SA. 2016. Ginger supplementation in nonalcoholic fatty liver disease: A randomized, double-blind, placebo-controlled pilot study. Hepatitis Monthly. **16**(1):34–89. DOI: 10.5812/hepatmon.34897

[14] Fauvelle C, Lambotin M, Heydmann L, Prakash E, Bhaskaran S, Vishwaraman M, Baumert TF, Moog C. 2017. A cinnamon-derived procyanidin type A compound inhibits hepatitis C virus cell entry. Hepatology International. **11**(5):440–445. DOI: 10.1007/s12072-017-9809-y

[15] Sarwar T, Almatroudi A, Almatroodi SA, Alharbi HOA, Rahmani AH. In *silico* analysis of bioactive compounds of *Nigella sativa* as potential inhibitors of NS5B RdRp protein involved in the pathogenesis of hepatitis C virus. Journal of Biomolecular Structure and Dynamics. 2025;1–20. DOI: 10.1080/07391102.2025.2515254

[16] Chen MH, Lee MY, Chuang JJ, Li YZ, Ning ST, Chen JC, Liu YW. 2012. Curcumin inhibits HCV replication by induction of heme oxygenase-1 and suppression of AKT. International Journal of Molecular Medicine. **30**(5):102–128. DOI: 10.3892/ijmm.2012.1096

[17] Tripathi AK, Ray AK, Mishra SK. Molecular and pharmacological aspects of piperine as a potential molecule for disease prevention and management: Evidence from clinical trials. Beni-Suef University Journal of Basic and Applied Sciences. 2022;**11**:16–37. DOI: 10.1186/s43088-022-00196-1

[18] Jiang ZY, Liu WF, Zhang XM, Luo J, Ma YB, Chen JJ. Anti-HBV active constituents from *Piper longum*. Bioorganic & Medicinal Chemistry Letters. 2013;**23**(7):21–45. DOI: 10.1016/j.bmcl.2013.01.118

[19] Takebe Y, Saucedo CJ, Lund G, Uenishi R, Hase S, Tsuchiura T, Kneteman N, Ramessar K, Tyrrell DLJ, Shirakura M, Wakita T. Antiviral lectins from red and blue-green algae show potent *in vitro* and *in vivo* activity against hepatitis C virus. PLoS One. 2013;**8**(5):64–89. DOI: 10.1371/journal.pone.0064449

[20] Pisonero-Vaquero S, García-Mediavilla MV, Jorquera F, Majano PL, Benet M, Jover R, González-Gallego J, Sánchez-Campos S. Modulation of PI3K-LXRα-dependent lipogenesis mediated by oxidative/nitrosative stress contributes to inhibition of HCV replication by quercetin. Laboratory Investigation. 2014;**94**(3):262–274. DOI: 10.1038/labinvest.2013.156

[21] Polyak SJ, Morishima C, Shuhart MC, Wang CC, Liu Y, Lee DYW. Inhibition of T-cell inflammatory cytokines, hepatocyte NF-κB signaling, and HCV infection by standardized silymarin. Gastroenterology. 2007;**132**(5):1925–1936. DOI: 10.1053/j.gastro.2007.02.038

[22] Wagoner J, Negash A, Kane OJ, Martinez LE, Nahmias Y, Bourne N, Owen DM, Grove J, Brimacombe C, McKeating JA, Pécheur EI. Multiple effects of silymarin on the hepatitis C virus lifecycle. Hepatology. 2010;**51**(6):1912–1921. DOI: 10.1002/hep.23587

[23] Neumann UP, Biermer M, Eurich D, Neuhaus P, Berg T. Successful prevention of hepatitis C virus (HCV) liver graft reinfection by silibinin

mono-therapy. Journal of Hepatology. 2010;**52**(6):951–952. DOI: 10.1016/j.jhep.2010.02.002

[24] Beinhardt S, Rasoul-Rockenschaub S, Scherzer TM, Ferenci P. Silibinin monotherapy prevents graft infection after orthotopic liver transplantation in a patient with chronic hepatitis C. Journal of Hepatology. 2011;**54**(3):591–592. DOI: 10.1016/j.jhep.2010.09.009

[25] Payer BA, Reiberger T, Rutter K, Beinhardt S, Staettermayer AF, Peck-Radosavljevic M, Ferenci P. Successful HCV eradication and inhibition of HIV replication by intravenous silibinin in an HIV–HCV coinfected patient. Journal of Clinical Virology. 2010;**49**(2):131–133. DOI: 10.1016/j.jcv.2010.07.006

[26] Calland N, Albecka A, Belouzard S, Wychowski C, Duverlie G, Descamps V, Hober D, Dubuisson J, Rouillé Y, Séron K. (-) Epigallocatechin-3-gallate is a new inhibitor of hepatitis C virus entry. Hepatology. 2012;**55**(3):720–729. DOI: 10.1002/hep.24803

[27] Chen C, Qiu H, Gong J, Liu Q, Xiao H, Chen XW, Sun BL, Yang RG. (-)-Epigallocatechin-3-gallate inhibits the replication cycle of hepatitis C virus. Archives of Virology. 2012;**157**(7): 1301–1312. DOI: 10.1007/s00705-012-1304-0

[28] Ciesek S, von Hahn T, Colpitts CC, Schang LM, Friesland M, Steinmann J, Manns MP, Ott M, Wedemeyer H, Meuleman P, Pietschmann T. The green tea polyphenol, epigallocatechin-3-gallate, inhibits hepatitis C virus entry. Hepatology. 2011;**54**(6):947–1955. DOI: 10.1002/hep.24610

[29] Haid S, Novodomská A, Gentzsch J, Grethe C, Geuenich S, Bankwitz D, Chhatwal P, Jannack B, Hennebelle T, Bailleul F, Keppler OT. A plant-derived flavonoid inhibits entry of all HCV genotypes into human hepatocytes. Gastroenterology. 2012;**143**(1):213–222. DOI: 10.1053/j.gastro.2012.03.036

[30] Calland N, Sahuc ME, Belouzard S, Pène V, Bonnafous P, Mesalam AA, Deloison G, Descamps V, Sahpaz S, Wychowski C, Lambert O. Polyphenols inhibit hepatitis C virus entry by a new mechanism of action. Journal of Virology. 2015;**89**(19):53–10063. DOI: 10.1128/jvi.01473-15

[31] Nahmias Y, Goldwasser J, Casali M, Van Poll D, Wakita T, Chung RT, Yarmush ML. Apolipoprotein B–dependent hepatitis C virus secretion is inhibited by the grapefruit flavonoid naringenin. Hepatology (Baltimore, Md.). 2008;**47**(5):1437–1445. DOI: 10.1002/hep.22197

[32] Goldwasser J, Cohen PY, Lin W, Kitsberg D, Balaguer P, Polyak SJ, Chung RT, Yarmush ML, Nahmias Y. Naringenin inhibits the assembly and long-term production of infectious hepatitis C virus particles through a PPAR-mediated mechanism. Journal of Hepatology. 2011;**55**(5):963–971. DOI: 10.1016/j.jhep.2011.02.011

[33] Choi M, Kim Y-M, Lee S, Chin Y-W, Lee C. Mangosteen xanthones suppress hepatitis C virus genome replication. Virus Genes. 2014;**49**:208–222. DOI: 10.1007/s11262-014-1098-0

[34] Tan WC, Jaganath IB, Manikam R, Sekaran SD. Evaluation of antiviral activities of Four Local Malaysian Phyllanthus species against herpes simplex viruses and possible antiviral target. International Journal of Medical

Sciences. 2013;**10**:1817–1829. DOI: 10.7150/ijms.6902

[35] Elsebai MF, Koutsoudakis G, Saludes V, Pérez-Vilaró G, Turpeinen A, Mattila S, Pirttilä AM, Fontaine-Vive F, Mehiri M, Meyerhans A, Diez J. Pan-genotypic hepatitis C virus inhibition by natural products derived from the wild Egyptian artichoke. Journal of Virology. 2016;**90**(4):1918–1930. DOI: 10.1128/jvi.02030-15

[36] Liu MM, Zhou L, He PL, Zhang YN, Zhou JY, Shen Q, Chen XW, Zuo JP, Li W, Ye DY. Discovery of flavonoid derivatives as anti-HCV agents *via* pharmacophore search combining molecular docking strategy. European Journal of Medicinal Chemistry. 2012;**52**:33–43. DOI: 10.1016/j.ejmech.2012.03.002

[37] Wu SF, Lin CK, Chuang YS, Chang FR, Tseng CK, Wu YC, Lee JC. Anti-hepatitis C virus activity of 3-hydroxy *caruilignan C* from Swietenia macrophylla stems. Journal of Viral Hepatitis. 2012;**19**(5):364–370. DOI: 10.1111/j.1365-2893.2011.01558.x

[38] Hassan ST, Berchová-Bímová K, Petráš J. Plumbagin, a plant-derived compound, exhibits antifungal combinatory effect with amphotericin B against *Candida albicans* clinical isolates and anti-hepatitis C virus activity. Phytotherapy Research. 2016;**30**(9):1487–1492. DOI: 10.1002/ptr.5650

[39] Lou S, Zheng YM, Liu SL, Qiu J, Han Q, Li N, Zhu Q, Zhang P, Yang C, Liu Z. Inhibition of hepatitis C virus replication *in vitro* by xanthohumol, a natural product present in hops. Planta Medica. 2014;**80**(02/03):171–176. DOI: 10.1055/s-0033-1360172

[40] Lan KH, Wang YW, Lee WP, Lan KL, Tseng SH, Hung LR, Yen SH, Lin HC, Lee SD. Multiple effects of Honokiol on the life cycle of hepatitis C virus. Liver International. 2012;**32**(6):989–997. DOI: 10.1111/j.1478-3231.2011.02621.x

[41] Hussein G, Miyashiro H, Nakamura N, Hattori M, Kakiuchi N, Shimotohno K. Inhibitory effects of Sudanese medicinal plant extracts on hepatitis C virus (HCV) protease. Phytotherapy Research. 2000;**14**:510–516. DOI: 10.1002/1099-1573(200011)14

[42] Shibata C, Ohno M, Otsuka M, Kishikawa T, Goto K, Muroyama R, Kato N, Yoshikawa T, Takata A, Koike K. The flavonoid apigenin inhibits hepatitis C virus replication by decreasing mature microRNA122 levels. Virology. 2014;**462-463**:42–48. DOI: 10.1016/j.virol.2014.05.024

[43] Wahyuni TS, Widyawaruyanti A, Lusida MI, Fuad A, Soetjipto, Fuchino H, Kawahara N, Hayashi Y, Aoki C, Hotta H. Inhibition of hepatitis C virus replication by chalepin and pseudane IX isolated from *Ruta angustifolia* leaves. Fitoterapia. 2014;**99**:276–283. DOI: 10.1016/j.fitote.2014.10.011

[44] Bachmetov L, Gal-Tanamy M, Shapira A, Vorobeychik M, Giterman-Galam T, Sathiyamoorthy P, Golan-Goldhirsh A, Benhar I, Tur-Kaspa R, Zemel R. Suppression of hepatitis C virus by the flavonoid quercetin is mediated by inhibition of NS3 protease activity. Journal of Viral Hepatitis. 2012;**19**(2):81–88. DOI: 10.1111/j.1365-2893.2011.01507.x

[45] Kong L, Li S, Liao Q, Zhang Y, Sun R, Zhu X, Zhang Q, Wang J, Wu X,

Fang X, Zhu Y. Oleanolic acid and ursolic acid: Novel hepatitis C virus antivirals that inhibit NS5B activity. Antiviral Research. 2013;**98**(1):44–53. DOI: 10.1016/j.antiviral.2013.02.003

[46] Adianti M, Aoki C, Komoto M, Deng L, Shoji I, Wahyuni TS, Lusida MI, Soetjipto, Fuchino H, Kawahara N, Hotta H. Anti-hepatitis C virus compounds obtained from *Glycyrrhiza uralensis* and other Glycyrrhiza species. Microbiology & Immunology. 2014;**58**:180–187. DOI: 10.1111/1348-0421.12127

[47] Wang L, Yang R, Yuan B, Liu Y, Liu C. The antiviral and antimicrobial activities of licorice, a widely-used Chinese herb. Acta Pharmaceutica Sinica B. 2015;**5**:310–315. DOI: 10.1016/j.apsb.2015.05.005

[48] Jardim AC, Igloi Z, Shimizu JF, Santos VA, Felippe LG, Mazzeu BF, Amako Y, Furlan M, Harris M, Rahal P. Natural compounds isolated from Brazilian plants are potent inhibitors of hepatitis C virus replication *in vitro*. Antiviral Research. 2015;**115**:39–47. DOI: 10.1016/j.antiviral.2014.12.018

[49] Wang L, Jiao H, Zhao J, Wang X, Sun S, Lin H. Allicin Alleviates Reticuloendotheliosis Virus-Induced Immunosuppression *via* ERK/Mitogen-Activated Protein Kinase Pathway in Specific Pathogen-Free Chickens. Frontiers in Immunology. 2017;**8**:1856. DOI: 10.3389/fimmu.2017.01856

[50] Hafid AF, Aoki-Utsubo C, Permanasari AA, Adianti M, Tumewu L, Widyawaruyanti A, Wahyuningsih SPA, Wahyuni TS, Lusida MI, Hotta H. Antiviral activity of the dichloromethane extracts from *Artocarpus heterophyllus* leaves against hepatitis C virus. Asian Pacific Journal of Tropical Biomedicine. 2017;**7**(7): 633–639. DOI: 10.1016/j.apjtb.2017.06.003

[51] Cagno V, Civra A, Kumar R, Pradhan S, Donalisio M, Sinha BN, Ghosh M, Lembo D. *Ficus religiosa L*. bark extracts inhibit human rhinovirus and respiratory syncytial virus infection *in vitro*. Journal of Ethnopharmacology. 2015;**176**:252–257. DOI: 10.1016/j.jep.2015.10.042

[52] Matsumoto Y, Matsuura T, Aoyagi H, Matsuda M, Hmwe SS, Date T, Watanabe N, Watashi K, Suzuki R, Ichinose S, Wake K. Antiviral activity of glycyrrhizin against hepatitis C virus *in vitro*. PloS One. 2013;**8**(7): 68–99. DOI: DOI: 10.1371/journal.pone.0068992

[53] Li LQ, Li J, Huang Y. Lignans from the heartwood of *Streblus asper* and their inhibiting activities to Hepatitis B virus. Fitoterapia. 2012;**83**(2):303–309. DOI: 1 0.1016/j.fitote.2011.11.008

[54] Wang HN, Yin ZF, Yin X, Li HB, Zhao GQ. Chemical constituents from *Pogonatherum crinitum* and their anti-HBV activities *in vitro*. Chinese Traditional Patent Medicine. 2019;**41**(6):234–256. DOI: 10.1155/2020/3972390

[55] Huang SX, Mou JF, Luo Q, Mo QH, Zhou XL, Huang X, Xu Q, Tan XD, Chen X, Liang CQ. Anti-hepatitis B virus activity of esculetin from *Microsorum fortunei in vitro* and *in vivo*. Molecules. 2019;**24**(19):34–75. DOI: 10.3390/molecules24193475

[56] Wang HL, Cao TW, Jiang FQ, Geng CA, Zhang XM, Huang XY, Wang LJ, Liang WJ, Rong GQ, Chen JJ.

Swerpunilactones A and B, the first example of xanthone and *secoiridoid heterodimers* from *Swertia punicea, S. hispidicalyx*, and *S. yunnanensis*. Tetrahedron Letters. 2013;**54**(21): 2710–2712. DOI: 10.1016/j. tetlet.2013.03.057

[57] Chen H, Li J, Wu Q, Niu XT, Tang MT, Guan XL, Li J, Yang RY, Deng SP, Su XJ. Anti-HBV activities of *Streblus asper* and constituents of its roots. Fitoterapia. 2012;**83**(4):643–649. DOI: 10.1016/j.fitote.2012.01.009

[58] Xu J, Gu W, Li C, Li X, Xing G, Li Y, Song Y, Zheng W. Epigallocatechin gallate inhibits hepatitis B virus via farnesoid X receptor alpha. Journal of Natural Medicines. 2016;**70**(3):584–591. DOI: 10.1007/s11418-016-0980-6

[59] Cao TW, Geng CA, Ma YB, He K, Wang HL, Zhou NJ, Zhang XM, Tao YD, Chen JJ. Xanthones with anti-hepatitis B virus activity from *Swertia mussotii*. Planta Medica. 2013;**79**(08):697–700. DOI: 10.1055/s-0032-1328399

[60] Zhang HL, Dai LH, Wu YH, Yu XP, Zhang YY, Guan RF, Liu T, Zhao J. Evaluation of hepatocyte protective and anti-hepatitis B virus properties of Cichoric acid from *Cichorium intybus* leaves in cell culture. Biological and Pharmaceutical Bulletin. 2014;**37**(7):1214–1220. DOI: 10.1248/bp b.b14-00137

[61] Geng CA, Yang TH, Huang XY, Yang J, Ma YB, Li TZ, Zhang XM, Chen JJ. Anti-hepatitis B virus effects of the traditional Chinese herb *Artemisia capillaris* and its active enynes. Journal of Ethnopharmacology. 2018;**224**:283–289. DOI: 10.1016/j.jep.2018.06.005

[62] Ge L, Wan H, Tang S, Chen H, Li J, Zhang K, Zhou B, Fei J, Wu S, Zeng X. Novel caffeoylquinic acid derivatives from *Lonicera japonica* Thunb. flower buds exert pronounced anti-HBV activities. RSC Advances. 2018;**8** (62):35374–35385. DOI: 10.1039/ c8ra07549b

[63] Cao TW, Geng CA, Ma YB, Zhang XM, Zhou J, Tao YD, Chen JJ. Chemical constituents of *Swertia mussotii* and their anti-hepatitis B virus activity. Fitoterapia. 2015;**102**:15–22. DOI: 10.1016/j.fitote.2015.01.020

[64] Cao TW, Geng CA, Ma YB, Zhang XM, Zhou J. Chemical constituents of *Swertia mussotii* and their anti-hepatitis B virus activity. Fitoterapia. 2015;**102**:15–22. DOI: 10.10 16/j.fitote.2015.01.020

[65] Jiang ZY, Liu WF, Zhang XM, Luo J, Ma YB, Chen JJ. Anti-HBV active constituents from *Piper Longum*. Bioorganic & Medicinal Chemistry Letters. 2013;**23**:2123–2127. DOI: 10.101 6/j.bmcl.2013.01.118

[66] Geng CA, Huang XY, Chen XL, Ma YB, Rong GQ, Zhao Y, Zhang XM, Chen JJ. Three new anti-HBV active constituents from the traditional Chinese herb of Yin-Chen (*Artemisia scoparia*). Journal of Ethnopharmacology. 2015;**176**:109–117. DOI: 10.1016/j.jep.2015.10.032

[67] Tsukamoto Y, Ikeda S, Uwai K, Taguchi R, Chayama K, Sakaguchi T, Narita R, Yao WL, Takeuchi F, Otakaki Y, Watashi K. Rosmarinic acid is a novel inhibitor for Hepatitis B virus replication targeting viral epsilon RNA-polymerase interaction. PLoS One. 2018;**13**(5):234–256. DOI: 10.1371/ journal.pone.0197664

[68] Rao VB, Yeturu K. Possible Anti-viral Effects of Neem (*Azadirachta*

Indica) on Dengue Virus. bioRxiv. 2020. DOI: 10.1101/2020.04.29.069567

[69] Huang RL, Chen CC, Huang HL, Chang CG, Chen CF, Chang C, Hsieh MT. 2000. Anti-hepatitis B virus effects of wogonin isolated from *Scutellaria baicalensis*. Planta Medica. **66** (08):694–698. DOI: 10.1055/s-2000-9775

[70] Huang TJ, Liu SH, Kuo YC, Chen CW, Chou SC. Antiviral activity of chemical compound isolated from *Artemisia morrisonensis* against hepatitis B virus *in vitro*. Antiviral Research. 2014;**101**:97–104. DOI: 10.1016/j.antiviral.2013.11.007

[71] Xia C, Tang W, Geng P, Zhu H, Zhou W, Huang H, Zhou P, Shi X. Baicalin down-regulating hepatitis B virus transcription depends on the liver-specific HNF4α-HNF1α axis. Toxicology and Applied Pharmacology. 2020;**403**:115–131. DOI: 10.1016/j.taap.2020.115131

[72] Chiang LC, Ng LT, Cheng PW, Chiang W, Lin CC. Antiviral activities of extracts and selected pure constituents of *Ocimum basilicum*. Clinical and Experimental Pharmacology and Physiology. 2005;**32** (10):811–816. DOI: 10.1111/j.1440-1681.2005.04270.x

[73] Shen B, Wu N, Shen C, Zhang F, Wu Y, Xu P, Zhang L, Wu W, Lu Y, Han J, Wang Y. Hyperoside nanocrystals for HBV treatment: Process optimization, *in vitro* and *in vivo* evaluation. Drug Development and Industrial Pharmacy. 2016;**42**(11): 1772–1781. DOI: 10.3109/03639045.2016.1173051

[74] Wu L, Yang X, Huang Z, Liu H, Sinica G. *In vivo* and *in vitro* antiviral activity of hyperoside extracted from *Abelmoschus manihot (L)*. Acta Pharmacologica Sinica. 2007;**28**: 404–409. DOI: 10.1111/j.1745-7254.2007.00510.x

[75] Guo Q, Zhao L, You Q, Yang Y, Gu H, Song G, Lu N, Xin J. Anti-hepatitis B virus activity of wogonin *in vitro* and *in vivo*. Antiviral Research. 2007;**74**(1):16–24. DOI: 10.1016/j.antiviral.2007.01.002

[76] Rouf R, Uddin SJ, Sarker DK, Islam MT, Ali ES, Shilpi JA, Nahar L, Tiralongo E, Sarker SD. Antiviral potential of garlic (*Allium sativum*) and its organosulfur compounds: A systematic update of pre-clinical and clinical data. Trends in Food Science and Technology. 2020;**104**:219–234. DOI: 10.1016/j.tifs.2020.08.006

[77] Kim HJ, Yoo HS, Kim JC, Park CS, Choi MS, Kim M, Choi H, Min JS, Kim YS, Yoon SW, Ahn JK. Antiviral effect of *Curcuma longa Linn* extract against hepatitis B virus replication. Journal of Ethnopharmacology. 2009;**124**(2):189–196. DOI: 10.1016/j.jep.2009.04.046

[78] Antoni S, Ferlay J, Soerjomataram I, Znaor A, Jemal A, Bray F. Bladder Cancer Incidence and Mortality: A Global Overview and Recent Trends. European Urology. 2017;**71**(1):96–108. DOI: 10.1016/j.eururo.2016.06.010

[79] Hartke J, Johnson M, Ghabril M. The Diagnosis and Treatment of Hepatocellular Carcinoma. Seminars in Diagnostic Pathology. 2017;**34**:153–159. DOI: 10.1053/j.semdp.2016.12.011

[80] Norikura T, Fujiwara K, Narita T, Yamaguchi S, Morinaga Y, Iwai K, Matsue H. Anticancer activities of

thelephantin O and vialinin A isolated from *Thelephora aurantiotincta*. Journal of Agricultural and Food Chemistry. 2011;**59**(13):6974–6979. DOI: 10.1021/jf200461j

[81] Kim C, Kim B. Anti-cancer natural products and their bioactive compounds inducing ER stress-mediated apoptosis: A review. Nutrients. 2018;**10**(8):10–21. DOI: 10.3390/nu10081021

[82] Wong JH, Ng TB, Chan HHL, Liu Q, Man GCW, Zhang CZ, Guan S, Ng CCW, Fang EF, Wang H, Liu F. Mushroom extracts and compounds with suppressive action on breast cancer: Evidence from studies using cultured cancer cells, tumor-bearing animals, and clinical trials. Applied Microbiology and Biotechnology. 2020;**104**(11):4675–4703. DOI: 10.1007/s00253-020-10476-4

[83] Xiao J, Liong EC, Ching YP, Chang RC, So KF, Fung ML, Tipoe GL. *Lycium barbarum* polysaccharides protect mice liver from carbon tetrachloride-induced oxidative stress and necroinflammation. Journal of Ethnopharmacology. 2012;**139**(2):462–470. DOI: 10.1016/j.jep.2011.11.033

[84] Tee TT, Cheah YH, Meenakshii N, Sharom MYM, Hawariah LPA. Xanthorrhizol induced DNA fragmentation in HepG2 cells involving Bcl-2 family proteins. Biochemical and Biophysical Research Communications. 2012;**420**(4):834–838. DOI: 10.1016/j.bbrc.2012.03.083

[85] Shimizu M, Shirakami Y, Sakai H, Tatebe H, Nakagawa T, Hara Y, Weinstein IB, Moriwaki H. EGCG inhibits activation of the insulin-like growth factor (IGF)/IGF-1 receptor axis in human hepatocellular carcinoma cells. Cancer Letters. 2008;**262**(1):10–18. DOI: 10.1016/j.canlet.2007.11.026

[86] Wang N, Feng Y, Zhu M, Tsang CM, Man K, Tong Y, Tsao SW. Berberine induces autophagic cell death and mitochondrial apoptosis in liver cancer cells: The cellular mechanism. Journal of Cellular Biochemistry. 2010;**111**(6):1426–1436. DOI: 10.1002/jcb.22869

[87] Zhang XL, Yu H, Xiong YY, Ma ST, Zhao L, She SF. Resveratrol down-regulates Myosin light chain kinase, induces apoptosis and inhibits diethylnitrosamine-induced liver tumorigenesis in rats. International Journal of Molecular Science. 2013;**14**(1):1940–1951. DOI: 10.3390/ijms14011940

[88] Abdel-Hamid NM, Nazmy MH, Nazmy WH. Vitamin C and diallyl sulfide as chemosensitizers to cisplatin in treating hepatocellular carcinoma. Journal of Solid Tumors. 2011;**1**(3):90–165. DOI: 10.5430/jst.v1n3p90

[89] Chen CY, Liu TZ, Liu YW, Tseng WC, Liu RH, Lu FJ, Lin YS, Kuo SH, Chen CH. 6-shogaol (alkanone from ginger) induces apoptotic cell death of human hepatoma p53 mutant Mahlavu subline via an oxidative stress-mediated caspase-dependent mechanism. Journal of Agricultural and Food Chemistry. 2007;**55**(3):948–954. DOI: 10.1021/jf0624594

[90] Gayathri R, Priya DK, Gunassekaran GR, Sakthisekaran D. Ursolic acid attenuates oxidative stress-mediated hepatocellular carcinoma induction by diethylnitrosamine in male Wistar rats. Asian Pacific Journal of Cancer Prevention : APJCP. 2009;**10**:933–938

[91] Mao WC, Song YJ, Zhang J, Jiao YB, Feng LL, Bai HF, Wang XP, Cai DY, Wang YQ. Effect of ursolic acid on DEN-induced hepatic precancerous lesions in mice. Chinese Journal of Integrated Traditional and Western Medicine on Liver Diseases. 2012;**22**:287–292

[92] Chuang SE, Kuo ML, Hsu CH, Chen CR, Lin JK, Lai GM, Hsieh CY, Cheng AL. Curcumin-containing diet inhibits diethylnitrosamine-induced murine hepatocarcinogenesis. Carcinogenesis. 2000;**21**:331–335. DOI: 10.1093/carcin/21.2.331

[93] Harrington K, Freeman DJ, Kelly B, Harper J, Soria J-C. Optimizing oncolytic virotherapy in cancer treatment. Nature Reviews Drug Discovery. 2019;**18**:689–706. DOI: 10.1038/s41573-019-0029-0

[94] Wu YY, Zhu L, Ma XY, Z.j S, Chen J, Chen XJ, Wan LH, Zhou LM. The anti-proliferation effect of Aikete injection on hepatocellular carcinoma *in vitro* and *in vivo*. Pharmaceutical Biology. 2011;**49**:531–538. DOI: 10.1007/s13205-023-03594-z

[95] Ji T, Lin C, Krill LS, Eskander R, Guo Y, Zi X, Hoang BH. Flavokawain B, a kava chalcone, inhibits growth of human osteosarcoma cells through G2/M cell cycle arrest and apoptosis. Molecular Cancer. 2013;**12**(1):55–98. DOI: 10.1186/1476-4598-12-55

[96] Zhang Z, Miao L, Lv C, Sun H, Wei S, Wang B, Huang C, Jiao B. Wentilactone B induces G2/M phase arrest and apoptosis *via* the Ras/Raf/MAPK signaling pathway in human hepatoma SMMC-7721 cells. Cell Death and Disease. 2013;**4**(6):657–687. DOI: 10.1038/cddis.2013.182

[97] Kanapathipillai M. Treating P53 Mutant Aggregation-Associated Cancer. Cancers. 2018;**10**(6):154–173. DOI: 10.3390/cancers10060154

[98] Chu Y-L, Ho C-T, Chung J-G, Raghu R, Lo Y-C, Sheen L-Y. Allicin induces anti-human liver cancer cells through the P53 gene modulating apoptosis and autophagy. Journal of Agricultural and Food Chemistry. 2013;**61**:9839–9848. DOI: 10.1021/jf403241s

[99] Liu M, Zhou J, Liu X, Feng Y, Yang W, Wu F, Cheung OKW, Sun H, Zeng X, Tang W, Mok MT. Targeting monocyte-intrinsic enhancer reprogramming improves immunotherapy efficacy in hepatocellular carcinoma. Gut. 2020;**69**:365–379. DOI: 10.1136/gutjnl-2018-317257

[100] Chen X, Hu X, Hu J, Qiu Z, Yuan M, Zheng G. Celastrol-loaded galactosylated liposomes effectively inhibit AKT/c-met triggered rapid hepatocarcinogenesis in mice. Molecular Pharmaceutics. 2020;**17**:738–747. DOI: 10.1021/acs.molpharmaceut.9b00428

[101] Sun H, Hou H, Lu P, Zhang L, Zhao F, Ge C, Wang T, Yao M, Li J. Isocorydine inhibits cell proliferation in hepatocellular carcinoma cell lines by inducing G2/M cell cycle arrest and apoptosis. PloS One. 2012;**7**(5):68–88. DOI: 10.1371/journal.pone.0036808

[102] Zanjirband M, Rahgozar S. Targeting P53-MDM2 interaction using small molecule inhibitors and the challenges needed to be addressed. Current Drug Targets. 2019;**20**: 1091–1111. DOI: 10.2174/1389450120666190402120701

[103] Yu H-Y, Zhang X-Q, Li X, Zeng F-B, Ruan H-L. 2-Methoxyjuglone induces apoptosis in hepg2 human hepatocellular carcinoma cells and exhibits *in vivo* antitumor activity in a h22 mouse hepatocellular carcinoma model. Journal of Natural Products. 2013;**76**:889–895. DOI: 10.1021/np400025b

[104] Su L, Chen X, Wu J, Lin B, Zhang H, Lan L, Luo H. Galangin inhibits proliferation of hepatocellular carcinoma cells by inducing endoplasmic reticulum stress. Food and Chemical Toxicology. 2013;**62**:810–816. DOI: 10.1016/j.fct.2013.10.019

[105] Chen W-T, Lin G-B, Lin S-H, Lu C-H, Hsieh C-H, Ma B-L, Chao C-Y. Static magnetic field enhances the anticancer efficacy of capsaicin on HepG2 cells *via* capsaicin receptor TRPV1. PLoS ONE. 2018;**13**:119–178. DOI: 10.1371/journal.pone.0191078

[106] Zhao Z, Jin G, Ge Y, Guo Z. Naringenin inhibits migration of breast cancer cells *via* inflammatory and apoptosis cell signaling pathways. Inflammo Pharmacology. 2019;**27**:10–21. DOI: 10.1007/s10787-01800556-3

[107] Rahman M, Almalki WH, Afzal O, Alfawaz Altamimi AS, Kazmi I, Al-Abbasi FA, Choudhry H, Alenezi SK, Barkat MA, Beg S, Kumar V. Cationic solid lipid nanoparticles of resveratrol for hepatocellular carcinoma treatment: Systematic optimization, *in vitro* characterization and preclinical investigation. International Journal of Nanomedicine. 2020;**15**:9283–9299. DOI: 10.2147/IJN.S277545

[108] Liu X, Du Q, Tian C, Tang M, Jiang Y, Wang Y, Cao Y, Wang Z, Wang Z, Yang J, Li Y. Discovery of CAPE derivatives as dual EGFR and CSK inhibitors with anticancer activity in a murine model of hepatocellular carcinoma. Bioorganic Chemistry. 2021;**107**:104–136. DOI: 10.1016/j.bioorg.2020.104536

[109] Sarfraz A, Rasul A, Sarfraz I, Shah MA, Hussain G, Shafiq N, Masood M, Adem Ş, Sarker SD, Li X. Hispolon: A natural polyphenol and emerging cancer killer by multiple cellular signaling pathways. Environmental Research. 2020;**190**:110–127. DOI: 10.1016/j.envres.2020.110017

[110] Zhao B, Ma Y, Xu Z, Wang J, Wang F, Wang D, Pan S, Wu Y, Pan H, Xu D, Liu L. Hydroxytyrosol, a natural molecule from olive oil, suppresses the growth of human hepatocellular carcinoma cells *via* inactivating AKT and nuclear factor-kappa B pathways. Cancer Letters. 2014;**347**(1):79–87. DOI: 10.1016/j.canlet.2014.01.028

[111] Wu N, Luo J, Jiang B, Wang L, Wang S, Wang C, Fu C, Li J, Shi D. Marine bromophenol bis (2, 3-dibromo-4, 5-dihydroxy-phenyl)-methane inhibits the proliferation, migration, and invasion of hepatocellular carcinoma cells *via* modulating β1-integrin/FAK signaling. Marine Drugs. 2015;**13**(2):1010–1025. DOI: 10.3390/md13021010

[112] Zhou M, Deng Y, Liu M, Liao L, Dai X, Guo C, Zhao X, He L, Peng C, Li Y. The pharmacological activity of berberine, a review for liver protection. European Journal of Pharmacology. 2021;**890**:173–255. DOI: 10.1016/j.ejphar.2020.173655

[113] Zhou R, Wu K, Su M, Li R. bioinformatic and experimental data decipher the pharmacological targets and mechanisms of plumbagin against

hepatocellular carcinoma. Environmental Toxicology and Pharmacology. 2019;**70**:103–200. DOI: 10.1016/j.etap.2019.103200

[114] White E, Lattime EC, Guo JY. autophagy regulates stress responses, metabolism, and anticancer immunity. Trends in Cancer. 2021;7:778–789. DOI: 10.1016/j.trecan.2021.05.003

[115] Mondal J, Khuda-Bukhsh AR. Cisplatin and farnesol co-encapsulated PLGA nano-particles demonstrate enhanced anti-cancer potential against hepatocellular carcinoma cells *in vitro*. Molecular Biology Reports. 2020;**47**:3615–3628. DOI: 10.1007/s11033-020-05455-x

[116] Li T, Xu W-S, Wu G-S, Chen X-P, Wang Y-T, Lu -J-J. Platycodin D induces apoptosis, and inhibits adhesion, migration and invasion in HepG2 hepatocellular carcinoma cells. Asian Pacific Journal of Cancer Prevention. 2014;**15**:1745–1779. DOI: 10.7314/APJCP.2014.15.4.1745

[117] Sarvizadeh M, Hasanpour O, Naderi Ghale-Noie Z, Mollazadeh S, Rezaei M, Pourghadamyari H, Masoud Khooy M, Aschner M, Khan H, Rezaei N, Shojaie L. Allicin and digestive system cancers: From chemical structure to its therapeutic opportunities. Frontiers in Oncology. 2021;**11**:50–86. DOI: 10.3389/fonc.2021.650256

[118] Mylonis I, Lakka A, Tsakalof A, Simos G. The dietary flavonoid kaempferol effectively inhibits HIF-1 activity and hepatoma cancer cell viability under hypoxic conditions. Biochemical and Biophysical Research Communications. 2010;**398**:74–78. DOI: 10.1016/j.bbrc.2010.06.038

[119] Shao J, Meng Q, Li Y. The flavins suppress tumor growth and metastasis *via* the blockage of the STAT3 pathway in hepatocellular carcinoma. Onco Targets and Therapy. 2016;**9**:4265–4275. DOI: 10.2147/OTT.S102858

[120] Sarika PR, James NR, Kumar PRA, Raj DK, Kumary TV. Gum Arabic-curcumin conjugate micelles with enhanced loading for curcumin delivery to hepatocarcinoma cells. Carbohydrate Polymers. 2015;**134**:167–174. DOI: 10.1016/j.carbpol.2015.07.068

[121] Wang Z, Zhang H, Zhou J, Zhang X, Chen L, Chen K, Huang Z. Eriocitrin from lemon suppresses the proliferation of human hepatocellular carcinoma cells through inducing apoptosis and arresting cell cycle. Cancer Chemotherapy and Pharmacology. 2016;**78**:1143–1150. DOI: 10.1007/s00280-016-3171-y

[122] Yang J, Pei H, Luo H, Fu A, Yang H, Hu J, Zhao C, Chai L, Chen X, Shao X, Wang C. Non-toxic dose of liposomal honokiol suppresses metastasis of hepatocellular carcinoma through destabilizing EGFR and inhibiting the downstream pathways. Oncotarget. 2016;**8**(1):915–945. DOI: 10.18632/oncotarget.13687

[123] Shimizu M, Shirakami Y, Sakai H, Tatebe H, Nakagawa T, Hara Y, Weinstein IB, Moriwaki H. EGCG inhibits activation of the insulin-like growth factor (IGF)/IGF-1 receptor axis in human hepatocellular carcinoma cells. Cancer Letters. 2008;**262**:10–18. DOI: 10.1016/j.canlet.2007.11.026

[124] Liang RR, Zhang S, Qi JA, Wang ZD, Li J, Liu PJ, Huang C, Le XF, Yang J, Li ZF. Preferential inhibition of hepatocellular carcinoma by the

flavonoid Baicalein through blocking MEK-ERK signaling. International Journal of Oncology. 2012;**41**:969–978. DOI: 10.3892/ijo.2012.1510

[125] Sun S, Shang E, Ju A, Li Y, Wu Q, Li Q, Yang Y, Guo Y, Yang D, Lv S. Tumor-targeted hyaluronic acid-mpeg modified nanostructured lipid carriers for cantharidin delivery: An *in vivo* and *in vitro* study. Fitoterapia. 2021;**155**: 105–133. DOI: 10.1016/j.fitote.2021. 105033

[126] Kunjiappan S, Sankaranarayanan M, Karan Kumar B, Pavadai P, Babkiewicz E, Maszczyk P, Glodkowska-Mrowka E, Arunachalam S, Ram Kumar Pandian S, Ravishankar V. Capsaicin-loaded solid lipid nanoparticles: Design, biodistribution, in silico modeling and *in vitro* cytotoxicity evaluation. Nanotechnology. 2021;**32**:95–101. DOI: 10.1088/1361-6528/abc57e

[127] Ko CY, Shih PC, Huang PW, Lee YH, Chen YF, Tai MH, Liu CH, Wen ZH, Kuo HM. Sinularin, an anti-cancer agent causing mitochondria-modulated apoptosis and cytoskeleton disruption in human hepatocellular carcinoma. International Journal of Molecular Sciences. 2021;**22** (8):39–46. DOI: 10.3390/ijms22083946

[128] Li D, Liu S, Zhu J, Shen L, Zhang QY, Zhu H. Folic acid modified TPGS as a novel nano-micelle for delivery of nitidine chloride to improve apoptosis induction in huh7 human hepatocellular carcinoma. BMC Pharmacology and Toxicology. 2021;**22**:10–34

[129] Zhang W-G, Yin X-C, Liu X-F, Meng K-W, Tang K, Huang F-L, Xu G, Gao J. Puerarin induces hepatocellular carcinoma cell apoptosis modulated by MAPK signaling pathways in a dose-dependent manner. Anticancer Research. 2017;**37**:4425–4431

[130] Bhattacharya S, Mondal L, Mukherjee B, Dutta L, Ehsan I, Debnath MC, Gaonkar RH, Pal MM, Majumdar S. Apigenin loaded nanoparticle delayed development of hepatocellular carcinoma in rats. Nanomedicine: Nanotechnology, Biology and Medicine. 2018;**14**: 1905–1917. DOI: 10.1016/ j.nano.2018.05.011

[131] Hu X, Zhang J, Deng L, Hu H, Hu J, Zheng G. Galactose-modified PH-sensitive niosomes for controlled release and hepatocellular carcinoma target delivery of tanshinone IIA. AAPS PharmSciTech. 2021;**22**:96–123. DOI: 10 .1208/s12249-021-01973-4

[132] Mandlik DS, Mandlik SK. Herbal and natural dietary products: Upcoming therapeutic approach for prevention and treatment of hepatocellular carcinoma. Nutrition and Cancer. 2021;**73**(11-12):2130–2154. DOI: 10.108 0/01635581.2020.1834591

[133] Batool S, Asim L, Qureshi FR, Masood A, Mushtaq M, Saleem RSZ. Molecular targets of plant-based alkaloids and polyphenolics in liver and breast cancer-an insight into anticancer drug development. Anti-Cancer Agents in Medicinal Chemistry. 2025;**25** (5):295–312. DOI: 10.2174/01187152 06302216240628072554

[134] Juaid N, Amin A, Abdalla A, Reese K, Alamri Z, Moulay M, Abdu S, Miled N. Anti-hepatocellular carcinoma biomolecules: Molecular targets insights. International Journal of Molecular Sciences. 2021;**22**(19): 107–174. DOI: 10.3390/ ijms221910774

[135] Wang Y, Li J, Xia L. Plant-derived natural products and combination therapy in liver cancer. Frontiers in Oncology. 2023;**13**:111–153. DOI: 10.3389/fonc.2023.1116532

[136] Maćków M, Dziubyna T, Jamer T, Slivinskyi D, Pytrus T, Neubauer K, Zwolińska-Wcisło M, Stawarski A, Piotrowska E, Nowacki D. The role of dietary ingredients and herbs in the prevention of non-communicable chronic liver disease. Nutrients. 2024;**16**(20):35–75. DOI: 10.3390/nu16203505

[137] Camilleri E, Blundell R. A comprehensive review of the phytochemicals, health benefits, pharmacological safety and medicinal prospects of *Moringa oleifera*. Heliyon. 2024;**10**(6):23–54. DOI: 10.1016/j.heliyon.2024.e27807

[138] Al-Toukhy GM, Suef RA, Hassan S, Farag MM, El-Tayeb TA, Mansour MT. Photobiological modulation of hepatoma cell lines and hepatitis B subviral particles secretion in response to 650 nm low level laser treatment. Journal of the Egyptian National Cancer Institute. 2023;**35**(1):33. DOI: 10.1186/s43046-023-00190-3

[139] Bhuker S, Kaur A, Rajauria K, Tuli HS, Saini AK, Saini RV, Gupta M. Allicin: A promising modulator of apoptosis and survival signaling in cancer. Medical Oncology. 2024;**41**(9):210. DOI: 10.1007/s12032-024-02459-6

[140] Asoka AS, Kolikkandy A, Nair B, Kamath AJ, Sethi G, Nath LR. Role of culinary Indian spices in the regulation of TGF-β signaling pathway in inflammation-induced liver cancer. Molecular Nutrition and Food Research. 2024;**68**(11):230–293. DOI: 10.1002/mnfr.202300793

[141] Pathak K, Pathak MP, Saikia R, Gogoi U, Sahariah JJ, Das A, Ahmad MZ, Paul T, Das J, Alqahtani SAM. Cardioprotective activities of some Indian spices: An insight into pharmacology and phytochemical investigation. Current Traditional Medicine. 2024;**10**(1):52–66. DOI: 10.2174/2215083809666230210105442

[142] Mandlik DS, Mandlik SK. Herbal and natural dietary products: Upcoming therapeutic approach for prevention and treatment of hepatocellular carcinoma. Nutrition and Cancer. 2021;**73**(11):2130–2154. DOI: 10.1080/01635581.2020.1834591

[143] Gupta M, Sarwat M. Protective effects of plant-derived natural products against hepatocellular carcinoma. Herbal Medicines. 2022;609–627. DOI: 10.1016/B978-0-323-90572-5.00009-3